TEXAS NATIONAL ENERGY MODELING PROJECT

AN EXPERIENCE IN LARGE-SCALE MODEL TRANSFER AND EVALUATION

TEXAS NATIONAL ENERGY MODELING PROJECT

AN EXPERIENCE IN LARGE-SCALE MODEL TRANSFER AND EVALUATION

Edited by

Milton L. Holloway
Energy Analysis and Development Division
Texas Energy and Natural Resource Advisory Council
Austin, Texas

ACADEMIC PRESS
A Subsidiary of Harcourt Brace Jovanovich, Publishers
New York London Toronto Sydney San Francisco
1980

ACADEMIC PRESS, INC.
111 Fifth Avenue, New York, New York 10003

United Kingdom Edition published by
ACADEMIC PRESS, INC. (LONDON) LTD.
24/28 Oval Road, London NW1 7DX

Library of Congress Cataloging in Publication Data

Main entry under title:

Texas national energy modeling project.

Includes index.
1. Energy policy--Texas--Addressess, essays, lectures.
2. Energy policy--United States--Addressess, essays, lectures. I. Holloway, Milton L.
HD9502.U53T488 333.79'09764 80-14613
ISBN 0-12-352950-6

PRINTED IN THE UNITED STATES OF AMERICA

80 81 82 83 9 8 7 6 5 4 3 2 1

TABLE OF CONTENTS

CONTRIBUTORS

PROJECT DIRECTOR

Milton L. Holloway, Texas Energy Advisory Council, Austin, Texas

NATIONAL ADVISORY BOARD

John S. Bonner, Bonner & Moore Associates, Inc., Houston, Texas

C. West Churchman, School of Business, University of California at Berkeley, Berkeley, California

George B. Dantzig, Department of Operations Research, Stanford University, Stanford, California

A. N. Halter, Electric Power Research Institute, Palo Alto, California

William W. Hogan, John F. Kennedy School of Government, Harvard University, Cambridge, Massachusetts

Walt W. Rostow, Department of Economics, The University of Texas at Austin, Austin, Texas

Robert M. Thrall, Department of Quantitative Management Science, Rice University, Houston, Texas, Chairman of the National Advisory Board

David O. Wood, MIT Energy Lab, Massachusetts Institute of Technology, Cambridge, Massachusetts

ANALYSIS TEAM

L. Martin Baughman, Center for Energy Studies, The University of Texas at Austin, Austin, Texas

Carol Bradley, Texas Transportation Institute, Texas A&M University, College Station, Texas

Bob Davis, Department of Agricultural Economics, Texas Tech University, Lubbock, Texas

W. E. Galloway, Bureau of Economic Geology, The University of Texas at Austin, Austin, Texas

William L. Fisher, Bureau of Economic Geology, The University of Texas at Austin, Austin, Texas

Robert R. Hill, Department of Business Analysis and Research, Texas A&M University, College Station, Texas

James E. Jonish, Department of Economics, Texas Tech University, Lubbock, Texas

Hong Y. Lee, Department of Agricultural Economics, Texas Tech University, Lubbock, Texas

William F. McFarland, Texas Transportation Institute, Texas A&M University, College Station, Texas

Jim Moore, Data Processing Center, Texas A&M University, College Station, Texas

Robert A. Morton, Bureau of Economic Geology, The University of Texas at Austin, Austin, Texas

Peter S. Rose, Department of Finance, Texas A&M University, College Station, Texas

F. D. Singleton, Industry Studies Program, University of Houston, Houston, Texas

John C. Stone, Industry Studies Program, University of Houston, Houston, Texas

Frank A. Taylor, Industry Studies Program, University of Houston, Houston, Texas

Theodore Taylor, Department of Economics, Texas Tech University, Lubbock, Texas

Russell G. Thompson, Industry Studies Program, University of Houston, Houston, Texas

Lawrence K. Vanston, Center for Energy Studies, The University of Texas at Austin, Austin, Texas

PREFACE

Beneath the surface of the rather technical pages that follow lies a drama and adventure of some general significance. The purpose of this brief preface, written collectively by the Advisory Board to the Texas National Energy Modeling Project, is to try to capture something of that drama and adventure and to set the work in the context of some fundamental philosophical issues raised by the study.

As in all good drama, there is an element of conflict in this story. President Carter's National Energy Plan (NEP), presented to Congress on April 20, 1977, took on a character greatly different from that proposed by two previous administrations. Although apparently using the same computer models as previous administration proposals and analyses, the supporting analysis for NEP resulted in very different projections of the nation's energy supply, demand and import balance, with and without the NEP. Further, the NEP emphasized policies intended to stimulate conservation with relatively little attention to encouraging production, a radical change from previous administrations. Most prominent of the new policies proposed in the NEP was a ceiling price for natural gas in the heretofore unregulated intrastate market; a wellhead tax for oil which would, when added to the regulated wellhead price, capture for the federal treasury the difference between the current price and the world oil price set by the Organization of Petroleum Exporting Countries (OPEC); and a mandatory system of boiler fuel conversions for utilities and major fuel-using industries.

Energy experts in Texas examined these policies and the supporting projections with some care. They concluded that the plan's expectations for abating the growing foreign crude oil imports to the U.S. could not be achieved under the policies which focused on conservation while not facing up to the conditions required to achieve NEP's production goals. Therefore, they had grave reservations about the realism of the oil, gas, coal and nuclear energy production targets projected as a result of NEP policies.

In the normal way scholars proceed when differences arise, the Texas experts asked the Federal Government for the assumptions and calculations which had led to NEP projections for 1985, in particular those relating to natural gas and oil production. It was generally known that these were embedded in a computerized system of interacting models then called the Project Independence Evaluation System (PIES).

Many months passed without a satisfactory response from Washington. The Executive Director of the Texas Energy Advisory Council (TEAC), in consultation with Lieutenant Governor William P. Hobby, the Texas Office of State/Federal Relations in Washington, D.C., and the academic experts at The University of Texas and the University of Houston, decided to request that the Federal Government, under the Freedom of Information Act, turn over its energy model for full professional study and critical analysis.

The formal request was made during October 1977.

At just about this time the organization of the Department of Energy (DOE) was being put together following its creation by the 95th Congress (PL 95-91, August 3, 1977), including within it an Energy Information Administration (EIA). EIA was given a rather interesting mandate by Congress, calling for (1) improved procedures to ensure credible energy data and analyses and (2) independence from the policy making offices of DOE and the Administration. These provisions reflected the judgment of Congress that a supportable national energy policy required a flow of information as reliable and politically uncolored as fallible human beings of integrity could make it.

EIA did not take the matter of the Freedom of Information Request seriously. The Texas group considered the lack of a response to comply with both the spirit and the letter of the law a serious matter. The TEAC Executive Director brought the matter to the attention of the Council Chairman, Lieutenant Governor Hobby. An emergency meeting of the Council was called, a resolution to Secretary Schlesinger and to members of the Congressional energy committees was adopted, and a meeting was scheduled between the Secretary and Lieutenant Governor Hobby.

A major Texas evaluation exercise was then organized, both to transfer and to evaluate the model—a process described in Chapter 1 of this report. In this setting, EIA eventually decided to drop its quasi-adversary posture towards the probing from Texas and cooperate in the transfer of the model. During the course of the transfer and evaluation, the drama became something of a shared adventure.

The broad significance of the adventure is simply this. Computers permit mathematical calculations to be made at a rate hardly conceivable in earlier times. This capacity, in turn, permits the implications of certain theoretically assumed relations to be worked out following a procedure involving the interaction of many more variables than could be handled in the precomputer world. Results can then be compared with what has happened in the past, and projections can be made. In computer modeling variables must be defined and assumptions stated precisely; vast amounts of data must be collected and organized; the variables must be linked through equations and other quantitative relations, usually based on past experience; and the computer can then roll out the implications of these equations and quantitative relations in the form of impressive printouts.

This kind of business, for good or ill, goes on in many government departments as well as in the private business sector, universities, and research centers. In its logical essence, model-building is quite simple. In practice, it has become a highly specialized and complex profession.

So far as government in a democracy is concerned, the use of computerized models raises a fundamental question: How are those outside the government to have confidence that the results of this process are worth a damn? Members of the Congress, state officials, interested private groups and the average citizen all are vitally concerned that the equations and the assumed quantitative relations reflect reasonably sound, even if stylized, approximation of how parts of the active world actually work. But without some method for piercing the wall behind which these arcane modeling operations proceed, outsiders are confronted by an enigmatic black box. An answer comes out of the box—a Delphic oracle that speaks in numbers. What goes on inside the black box is hidden. A democracy cannot live comfortably with this kind of jiggery-pokery. Debate in a democracy requires that the assumptions underlying policy conclusions be rendered, in the end, explicit. It also requires that affected people be able to know and understand—know what kind of data and assumptions go in and by what logic or process conclusions are drawn; until they know these, models lack credibility with them.

Thoughtful modelers have been quite conscious of this problem. They have considered with much care how the models used in government can be set up also outside government; how their equations and quantitative relations should be examined and criticized; and what the areas of common ignorance, the issues for debate, and the tasks revealed for future research and study are. For the EIA, the Texas exercise involved, no doubt, a diversion of scarce time; but it was also a unique opportunity to validate the integrity of its modeling efforts.

A more important question of philosophy has been brought to the surface in this exercise. The idea that models and modelers should be totally isolated from interested users is an issue of great importance. The Texas group believes that to ignore feedback from interested parties and users is to lose a vital source of information, leading to irrelevance and ignorance on the part of modelers. While some insulation is certainly needed, isolation is ill advised. Deliberate avenues of interaction with users such as the Texas group need to be structured as an ongoing part of an agency's model development and model use program. Further, we believe that knowledge about values, non-monetary as well as monetary, can be handled objectively and explicitly, and is required. It is not objective to avoid the value questions which are at the heart of problems involving energy.

The reader should be aware that large-scale model evaluation and transfer is not a trivial task. For those in Texas who had forced the opening of this particular black box, success involved a sobering challenge with three dimensions.

First, experts had to be found capable of effecting the transfer of the model to a computer in Texas. This involved vastly more than mailing the computer tapes from Washington, D.C., to College Station where the Texas A&M computer is located. It required the tracking down of all manner of precise information which had not been well documented and filed in the midst of the day-to-day workings of a hard pressed Washington bureaucracy. Together the Texas and EIA computer experts managed to effect the transfer—a task complicated by the fact that the computers at the two places were not identical. It did not prove possible wholly to reproduce the model which underlies the NEP; but the model underlying the EIA April 1978 Administrator's Annual Report to the Congress has, to all intents and purposes, been successfully set up and used in Texas.

We would underline here a major substantive conclusion of this project: the transfer of a model is not an antiseptic mechanical matter; it requires sustained collegial cooperation at both ends among the most knowledgeable experts.

The second challenge facing the Texas team was to mobilize a group of experts capable of examining the equations and quantitative relationships assumed in the various components of the integrated EIA model. Here Texas was lucky. As a state deeply involved in the production and industrial processing of energy, Texas is rich in expertise of high academic competence. TEAC provides an excellent link between policymakers and academic researchers and further provides the leadership needed to form inter-university working groups. The eleven critical studies of component models included in this report were done, in their individual capacity, by experts at The University of Texas at Austin, Texas A&M University, the University of Houston, and Texas Tech University. Amidst their other responsibilities they found the time to write serious evaluations. They not only provided independent evaluations of the theoretical assumptions and quantitative relations used in Washington, but also constructive suggestions for their improvement. Moreover, although time was short, they have been able to use the model now operating in Texas to illuminate the significance for analysis and policy of some of the improvements they commend.

The third challenge facing the Texas team was to find a way to examine the intellectual integrity of their efforts in transfer, analysis and criticism. For these purposes, a National Advisory Board was recruited with a majority of its members from states other than Texas. In one way or another, its members all had some experience with the problems of modeling; and they were united in their judgment that it was important for the workings of a democratic society that models used inside government be replicated outside with their theoretical assumptions and assumed quantitative relationships revealed and subjected to critical analysis, and that an avenue of interaction between users and model developers be established.

Our task as members of the National Advisory Board was not to try to impose our substantive views on either the DOE or the Texas experts. Our task was to follow the processes of transfer and analysis; to vouch for the integrity of the effort; and to offer our suggestions as friendly observers on both the exercise described in this book and for future work.

In discussing the desirability of publishing the findings and conclusions of TNEMP, we considered several alternatives, ranging from the level of editorial review customary in journals of professional organizations to a project report with little attention to editorial niceties. We felt that the timeliness of the material would be lost if we waited for the full editorial treatment and hence opted for quick publication as is. Although the eleven studies in Part II, available under separate cover from the Texas Energy Advisory Council, have undergone considerable review and revision, they stand as finally submitted by their authors on their responsibility—not that of the National Advisory Board. Part I of the report has been reviewed more extensively.

As our report included in Chapter 4 indicates, we can attest that this was a good faith effort at both the Washington and Texas ends of the line; we believe a good many of the conclusions of the Texas analysts deserve extremely serious consideration and follow-up to clarify their importance; and we recommend that the breakthrough this book describes be converted into an ongoing process at the service not only of Texas but the nation as a whole.

When we accepted our responsibility as members of the National Advisory Board we believed the effort was worth making; but we were uncertain of its success. It is now our conclusion that the adventure succeeded beyond our expectations and, with all its imperfections, represents a breakthrough which should be enlarged in energy and other fields.

Members of the
National Advisory Board to the
Texas National Energy Modeling Project

READER'S GUIDE AND ACKNOWLEDGEMENTS

READER'S GUIDE

An evaluation of the Department of Energy's primary energy forecasting and policy analysis tool, the Mid-Term Energy Forecasting System, by an office of Texas state government will on the surface appear suspect to many readers, since Texas energy policy proposals have often been at odds with those of the federal government. For this reason a major effort was made to ensure credibility of the project by structuring input from and an assessment of the effort by a national professional group and to allow the opportunity of reply to the evaluation by the Department of Energy. Therefore, the reader will find in Chapter 4 the National Advisory Board assessment of the Texas National Energy Modeling Project (TNEMP). The reader will likewise find in Chapter 5 a reply by the Department of Energy.

In order to ensure that the project was not merely a staff exercise, the Chairman of the Texas Energy Advisory Council, Lieutenant Governor William P. Hobby, met with the National Advisory Board (NAB) twice during the course of the project; and the Administrator of the Energy Information Administration (EIA) of the Department of Energy assigned a liaison person, Dr. Elizabeth MacRae, Deputy Assistant Administrator for Applied Analysis, to the project for coordination between EIA and TNEMP. To achieve attention and discussion of the findings by top-level officials, a meeting was held at the end of the project between DOE officials, Lieutenant Governor Hobby, members of the National Advisory Board, and the Project Director who is also the Executive Director of the Texas Energy Advisory Council. All four major TNEMP recommendations were favorably received by DOE/EIA at the meeting during June 1979.

The reporting of the TNEMP experience has the following organization. The writing is divided into two parts in order to keep cost low for those not interested in some of the detail. Part I, published under this cover, contains all of the summary material documenting the entire experience, sequentially, from beginning to end. The focus of Part I is the important findings of the study.

An extensive preface was written for Part I by the National Advisory Board in order to put the TNEMP experience in perspective from their point of view and to clarify the Board's role in the entire affair. An acknowledgements section was written by the Project Director and lists those involved in TNEMP. A Highlights section summarizes the project for the convenience of the reader.

Chapter 1 lays out the purposes of TNEMP, the organizational structure for the study, and an explanation of the evaluation criteria used to guide the model critiques. Chapter 2 summarizes in some detail the important findings of each of the 11 studies contained in Part II published under a separate cover. The section reflects the author's assessments prior to any formal review and comment from EIA. Chapter 3 was written by the Project Director and lays out a suggested organizational structure for continued Texas work. Chapter 4 was written by the National Advisory Board and reflects their assessment of the integrity of the evaluation project, their views of important outcomes of the TNEMP experience, and important recommendations to TNEMP and EIA. Chapter 5 contains an overview reply by EIA; a more detailed review statement is published with Part II. Chapter 6 is a summary of a workshop held at the end of the project to discuss substantive issues raised by TNEMP. Appendix B contains abstracts of each of the 11 reports which are the subject matter of Part II. The interested reader will want to obtain Part II for the details leading to the summary statements in the abstracts. Part II is published under a separate cover and available from the Texas Energy Advisory Council, Executive Office Building, 411 West 13th Street, Austin, Texas 78701.

Chapters 1 through 4 and Appendix B of Part I and the 11 studies of Part II were written prior to any formal review and comment by the Energy Information Administration and have not been updated to reflect any changes in the views of the authors following the EIA review and the workshop held to discuss the issues. Chapter 5 was written as a formal review by EIA. More detailed review comments by EIA on the 11 studies contained in Part II are published with Part II since they relate directly to the individual studies. The EIA comments were written following the completion of Chapters 1 through 4 and the 11 studies of Part II, and do not reflect results of the workshop. Finally, Chapter 6 summarizes the results of a workshop held in Washington, D.C. at the end of the project. The workshop was held in order to clarify significant agreements and remaining disagreements between EIA and TNEMP and to identify needed further work.

Acknowledgements

Much credit for the completion of the TNEMP must be given to the members of the National Advisory Board. Without their advice, guidance and influence the project would not have succeeded. Dr. Robert M. Thrall of Rice University served ably as Chairman of the NAB, providing counsel to the Project Director throughout the project and presiding at National Advisory Board meetings. Other members, each providing unique perspectives and backgrounds, included Dr. C. West Churchman, University of California at Berkeley; Dr. George B. Dantzig, Stanford University; Dr. Albert Halter, Electric Power Research Institute; Mr. John Bonner, Bonner & Moore, Associates, Inc., Houston; Mr. David O. Wood, MIT Energy Lab; Dr. William W. Hogan, Harvard University; and Dr. Walt W. Rostow, University of Texas at Austin.

The Analysis Team completed various evaluations of selected parts of the Mid-Term Energy Forecasting System and provided a depth and breadth of expertise and experience perhaps unmatched in modeling evaluation elsewhere. The major accomplishment of the transfer and operation of the very large computer modeling system could not have been accomplished without the endless hours of effort from Dr. Jim Moore, Texas A&M University, who was never content with partial success and by his own initiative spent many hours conferring with EIA modelers and systems analysts to make the system operational and to complete sensitivity analyses for other principal investigators.

Dr. William Fisher, Director, Bureau of Economic Geology, The University of Texas at Austin, with Dr. Robert A. Morton and Dr. W. E. Galloway, also of the Bureau, provided expertise ideal for the purpose of examining the influence of reserve estimates and finding rates on oil and gas production projections. Dr. Martin Baughman, Director of the Electric Power Systems Division of the Center for Energy Studies at The University of Texas at Austin and developer of a nationwide regional electricity model, with Larry Vanston, also of the Center, provided the evaluation of the electric power portion of the MEFS.

Dr. Peter Rose, Finance Department, Texas A&M University, who has written broadly on the financial aspects of the oil and gas industry, provided the evaluation of the investment model of the oil and gas portion of MEFS. Dr. Robert R. Hill, College of Business Administration, Texas A&M University, and an early developer of a U.S. oil and gas supply model, provided an evaluation of the adequacy of the MEFS oil and gas model to represent the growing "stripper" well portion of the industry. Dr. W. Frank McFarland, Texas Transportation Institute, Texas A&M University, and author of many publications on the economics of transportation, with Carol Bradley, also of the Institute, provided an evaluation of the transportation sector of the MEFS.

Dr. Russell G. Thompson, Director of Industry Studies, and Mr. John Stone, Dr. F. D. Singleton, Jr. and Dr. Frank A. Taylor, Research Associates at the University of Houston and developers of their own oil and gas supply model, refinery model, and with prior experience in operation of the MEFS coal supply model, provided hands-on operation as well as evaluation of MEFS oil, gas, coal, and refinery sector models.

Dr. James E. Jonish, Chairman, Department of Economics, Texas Tech University, with experience in macroeconomic modeling, and Dr. Theodore Taylor, also of the Department, completed an evaluation of the MEFS/DRI interface, including the regional and international aspects of the modeling system. Dr. Hong Lee and Dr. Bob Davis, Agricultural Economics Department, Texas Tech University, both experienced in applied econometrics, provided an evaluation of the demand model portion of MEFS.

Early in the project it was recognized that various normative issues concerning the models, and the Washington and Texas institutions involved in the evaluation, should be studied. Dr. Glenn Johnson and Ms. Judy Brown of Michigan State University were eventually retained to complete such a study. Dr. Johnson has written extensively in the area of normative issues and has directed major modeling development projects. Although the resulting study is not included in this publication since it was completed late in the project, I am confident that it will be very useful in the ongoing work of TNEMP, and perhaps of EIA.

Several individuals representing institutions supporting the objectives of the study and with mutual interests in further developing Texas modeling for policy analyses, were organized for participation in TNEMP. The group is known for purposes of TNEMP as Supporting Institutions and includes Dr. Charles Holt, Director, Bureau of Business Research, The University of Texas at Austin; Dr. William Fisher, Director, Bureau of Economic Geology, The University of Texas at Austin; Dr. George Kozmetsky, Dean of the Graduate Business School, The University of Texas at Austin; Dr. Herbert Woodson, Director, Center for Energy Studies, The University of Texas at Austin; Dr. Spencer Baen, Director, Center for Energy and Mineral Resources, Texas A&M University; Dr. Alvin Hildebrandt, Director, The Energy Laboratory, University of Houston; Dr. Marion Hagler, Director, Center for Energy Research, Texas Tech University; and Dr. John Freeman, Director, Space Solar Power Research Program, Rice University. Finally, Texas Governor Dolph Briscoe provided the funding from the Governor's Office for completing this study.

Excellent support from the Texas Energy Advisory Council staff allowed the project director to attend to the business of directing the day-to-day affairs of the project—the National Advisory Board meetings, coordination among members of the Analysis Team, liaison with the Energy Information Administration; assigning priorities for work and communicating with the rather large group of interested participants—while carrying on the regular duties of the Executive Director of the Texas Energy Advisory Council. Mr. Larry Buch provided communication and reference material for Analysis Team members. Ms. Sheila Moritz provided meeting arrangements and helped with timely communication with project participants. Mr. Charles Mueller provided the accounting and fiscal affairs of contract negotiations and administration for the project. Mr. Edward Selig provided editorial assistance. Ms. Vickie Everhart provided typing and secretarial support for the Project Director during the course of the project.

Austin, Texas
August 1979

Dr. Milton L. Holloway
Project Director for the
Texas National Energy Modeling Project

LIST OF ACRONYMS

AAR .... Administrator's Annual Report to Congress
ACB .... Air Control Board
AEPR .... Annual Energy Policy Review
AT .... Analysis Team
black box .... mystical or hidden information
boe .... barrels of oil equivalent
CHRDS .... Comprehensive Human Resources Data System
CMT .... Core Modeling Team
DASD .... direct access storage devices
DD* data files .... appears as part of the submitted computer run
DOE .... Department of Energy
DRI .... Data Resources, Inc.
ECPA .... Energy Conservation and Production Act
EPCA .... Energy Policy and Conservation Act
EIA .... Energy Information Administration
IEES .... International Energy Evaluation System
I/O .... input/output
MAGEN .... Haverly Systems, Inc.'s matrix and report generator
mcf .... 1,000 cubic feet
MEFS .... Midterm Energy Forecasting System
MEMM .... Midterm Energy Market Model
MIT .... Massachusetts Institute of Technology
MPS3 .... Ketron, Inc.'s advanced high-performance mathematical programming system
MPSX/360 .... version 360 of IBM's mathematical programming system extended
MVS .... multiple virtual storage
NAB .... National Advisory Board
NCM .... National Coal Model
NEP .... National Energy Plan
NPV .... net present value
OMS .... Oil Market Simulation Model
PART .... Professional Audit Review Team
PIES .... Project Independence Evaluation System
PILOT .... energy economic model with process-type modeling in the energy sector
PRC .... policy research contractors
PRG .... Peer Review Group
RDFOR .... Regional Demand Forecasting Model
REPS .... Regional Emission Projection System
RG .... Research Group

SVS ....single virtual storage
TEAC ....Texas Energy Advisory Council
TEPP ....Texas Energy Policy Project
TNEMP...............Texas National Energy Modeling Project
USGS ....United States Geological Survey

HIGHLIGHTS

Several major study conclusions and recommendations stand out as important findings of the Texas National Energy Modeling Project (TNEMP) evaluation of the Department of Energy, Energy Information Administration's (DOE/EIA) Mid-Term Energy Forecasting System (MEFS).

OVERVIEW

—The TNEMP has achieved a major mechanical transfer of a recent version of the MEFS (formerly known as the Project Independence Evaluation System) maintained and used for energy projections and policy analyses by DOE/EIA (and formerly by the Federal Energy Administration). Major progress has also been made in the operation of the computer models by TNEMP evaluators to allow assessment of the model's structure, behavior, and usefulness for energy policy analyses. This model transfer could not have occurred nor could the limited evaluations requiring model operation on the computer have been accomplished without major cooperation from DOE/EIA. TNEMP represents a major step forward in improvement of model transfer and evaluation required to advance the credibility and usefulness of large-scale models for public policy analyses.

HISTORY

—The story of the transfer and evaluation of MEFS begins in 1973, both in Washington, D.C. and in Texas. In Washington there were groups in the Interior Department as well as the Office of Management and Budget who were considering the rudiments of the energy modeling system, even before the "crisis" of the embargo. After the embargo and the executive directive of "energy independence" of the U.S. by 1980, a group was formed in the new Federal Energy Office to make sense out of the available data on energy, which were generally regarded to be a "mess." After considering several alternatives, which included writing a 50-page "think piece," the group decided to put the pieces together by building a large-scale integrated model, which was called Project Independence Evaluation System (PIES). PIES made its appearance in a report in November, 1974, known as Project Independence Blueprint. In the appendix to the report, the authors admitted that there were serious deficiencies in the modeling and data systems; e.g., in the demand model and the way in which it was integrated into the rest of the PIES model. The authors fully expected that the enthusiasm which had driven them to the completion of the first PIES would continue and drive them into making the appropriate corrections. But bureaucracy took over as the energy issue became the central, first priority business of the nation. The Federal Energy Office, the Federal Energy Administration, and then DOE with its EIA were formed. The original group was enlarged, several of the authors of the report left and, in effect, the resolve as expressed in the appendix to the 1974 Project Independence Blueprint to mend the deficiencies, was lost. All that remained was PIES itself, in many respects poorly documented because the model builders in their enthusiasm and under time pressure, had not clearly documented their work and had in the meanwhile disappeared into that murky area called un-Washington, D.C. In 1978, perhaps as a symbolic act of recognition of its potential deficiencies, PIES was renamed Mid-Term Energy Forecasting System (MEFS) to distinguish its use and its management from that of the old organization and perhaps to emphasize that the model did not have long-range forecasting potential.

Meanwhile, back in Texas in 1973, there was created a government organization to deal with Texas energy and several attempts to model aspects of the energy system. A Texas Governor's Energy Advisory Council was formed. The Council was itself reorganized twice over the years and currently exists under the name of Texas Energy Advisory Council (TEAC), with the Lieutenant Governor as its Chairman and Milton Holloway as its Executive Director. Holloway had heard of PIES and knew about the "mysterious affair of the National Energy Plan," supported or justified by a set of projections with the plan and without it, but no explanation was given of the origin of the projections. After several months of inquiry, the existence of PIES as the analytical tool used was revealed. After several more months of frustrated attempts by experts in Texas to gain access to the data and models, even after filing a Freedom of Information Act request, TEAC decided to insist on the transfer of the model from Washington to Texas. Transferring a model is not like picking up a tape and mailing it. One has to understand how the model works, and this is a complicated process. TEAC also decided to evaluate the model under four general criteria discussed in greater detail in this report.

MAJOR FINDINGS

—Not surprisingly, given the history and setting described above, MEFS (PIES) did not measure up well under TNEMP evaluation criteria defined as (1) workability (aids in addressing important practical problems), (2) clarity (is relatively unambiguous for analysts and users), (3) validity (the components of the model interact among themselves in a manner logically consistent with theory or accepted understanding of how the system being studied really behaves), and (4) verifiability (there is some practical way to demonstrate that the model's consequences and behavior agree reasonably well with reality). Perhaps other large models, given the current state of the art, would not fare well under such criteria either, but we should not be content with the status quo.

—MEFS results are highly dependent upon the diffuse judgment of many modelers within DOE. This means that outsiders must necessarily have considerable faith in the integrity and judgment of the modelers themselves, as well as the "objective scientific" performance of a computerized modeling system.

—Given the current state of complexity and inadequate documentation of MEFS, transferability cannot be a mechanical operation relying on current written documentation. Transferability requires major interaction between the modelers and the model assessors or users of the transferred modeling system. The complexity of the modeling systems, the lack of clarity in the concepts and operations of the models, and the poor and/or non-existent documentation and almost total lack of information verification makes this kind of interaction absolutely essential to accomplish transfer; without transferability of the model and outside testing, credibility badly suffers.

—Major reforms are needed to increase the reliability of information from MEFS. More specifically, third party institutions should specify standards for model documentation and verification procedures. Still another third party institution should be responsible for periodic assessments of the model's overall reliability in terms of criteria similar to that used in this study. The current DOE/EIA plan for archival of MEFS at Argonne, though helpful, is not adequate to achieve transfer; a "living" model must be maintained.

THE MODELS

—The oil and gas supply models are seriously weak on several grounds and, as is presently planned, EIA would be wise to replace the model. The development of a new model, however, should make use of several findings of TNEMP. First, the representation of the investment behavior of the oil and gas industry is currently simplistic with no apparent empirical base. A survey of the recent economics and financial literature, as well as survey results from TNEMP, suggest that a more complex and empirically based investment model is needed. Second, the models treat oil and gas as separate products when in fact the two hydrocarbon sources are joint products in the production process. Third, the discount rate used in the model is only about one-half that currently representative for the industry and considered necessary to encourage investment. Fourth, the finding rate calculation in the model is arbitrarily defined as a smooth declining function of time constrained by the U.S. Geological Survey's estimates of reserves; empirical evidence suggests a functional form reflecting various plateaus as a function of time. Further research is needed to explain this observed behavior. Given the uncertainty of this parameter, variations should be provided in analyses since this factor is obviously of great importance in production projections. The behavior of the gas model shows very little production responsiveness for prices near current market levels, a model result which is empirically questionable. Verification of such model behavior is badly needed. Fifth, the model documentation is so poor and/or non-existent that it is virtually impossible for an outside user to operate the model with confidence. The usefulness of the model for practical purposes of policy analysis concerning wellhead price controls, tax incentives, wellhead taxes, and other policy options is therefore poor. The model lacks workability.

—The coal supply model is comparatively better than the oil and gas models on matters of clarity and related documentation. To meet the criterion of verifiability, more work on the model needs to be done since the results indicate a more rapid industry production response than observed empirically.

—The integrating model representations of the electric power, refinery, and transportation sectors are reasonably satisfactory given their limited purpose as "place holders" in the market equilibrium structure of MEFS; the overall solution of MEFS is influenced primarily by the supply and demand specifications. There are certain exceptions. For example, the refinery sector representation produces reasonable results (i.e., reasonable prices for the mix of refinery products) for only a rather narrow variation in product mixes thus limiting the usefulness of MEFS for some interesting analyses. The transportation model representation is weak on verification grounds when used for regional analyses.

—The demand model, which is central to MEFS projections of national energy consumption and estimates of the economy's ability to substitute fuels, does not contain certain test statistics used by econometricians to judge the reliability of the empirical parameter estimates. One can only guess at reliability by comparing results with other studies which do have available test statistics. Sensitivity tests to determine the importance of the reliability of econometrically estimated elasticity estimates in the overall model solutions were mostly unsuccessful. The procedures for completing such sensitivity tests operationally were not provided by EIA and apparently such tests have not been performed by EIA. Documentation of how to correctly complete sensitivity tests of the demand model parameters needs to be specified by EIA, and more importantly, EIA needs to understand the importance of variation in elasticity parameters on MEFS results.

—The Data Resources, Inc. (DRI) macroeconomic model used to drive MEFS is considered a "state of the art" model, and the idea of interfacing DRI with a detailed energy market model is an innovative approach. However, the linkage between the macro and micro systems is very simplistic and ignores important interrelationships. Major methodological problems arise because of attempts to integrate several modeling techniques. Major sectoral and regional shifts in energy-related investment resulting from changes in OPEC prices or U.S. government policy have virtually no impact on the macro model results. The effects of U.S. energy sector changes on world trade and its effect on the value of the dollar are not directly modeled. Thus, the overall modeling system can only provide partial information and could be misleading about the most frequently asked questions by energy policymakers: (1) What is the effect on U.S. economic growth of alternative energy policy programs? (2) How will rising crude oil imports affect the U.S. balance of trade, balance of payments, and value of the dollar?

—A major weakness of the modeling system concerns its usefulness for regional impact analyses. The methodology does not capture the regional shifts in economic activity driven by changing investment patterns resulting from changing relative prices of energy or government policy affecting the relative costs of doing business. Model results also indicate this weakness. For this reason the model is not usable for some of the most important policy questions facing policymakers—diverse and uneven regional impacts.

MODEL ASSESSMENT CRITERIA

—There is a need to further clarify a taxonomy of evaluation criteria. This exercise is a prerequisite to a common understanding of what measures should be used to test the reliability of information from large-scale models. This project has added to the understanding of what is involved in third-party model assessments and transfer.

Without DOE/EIA cooperation and many hours of personal communications at the staff level, transfer of MEFS, though still incomplete, could not have been accomplished. This fact reinforces our conclusion that transferability must be more than a mechanical operation.

CONTINUED TEXAS WORK

—Texas should further develop its own modeling capability for assessing the regional impacts of national energy policy proposals for input into the decision making process; no other institutional entity can be expected to adequately complete such analyses. Such a modeling effort should be done taking full advantage of the results of TNEMP's experience.

—Texas should continue to transfer current "living versions" of MEFS to Texas for operation and maintenance as a service function to other users, thus helping EIA to fulfill its mandate to provide access to modeling systems and data bases maintained for public policy work.

—Texas should institutionalize an annual evaluation of EIA's Administrator's Annual Report to Congress, reviewing alternative projections and outlook for the benefit of Texas and the nation.

CHAPTER 1
PROJECT PURPOSE AND STRUCTURE

by Milton L. Holloway

INTRODUCTION

Large-scale models are a product of our times and seem to be growing in the importance of their use in government policy work. Policymakers are using large-scale models, having no alternative due to the complexity of potential impacts from policy decisions, but at the same time they are skeptical of the reliability of forecasts and calculations from models. There is also great acceptance by the populace at large of results coming from computer analyses which in their minds seem to represent the epitome of technological solutions to problems. Analyses are somehow seen as more believable if they are based on computer technology.

A recent article in *Fortune* characterizes the importance of this relatively new modeling phenomena in policy work.[1]

> "When the history of economic policymaking in the turbulent late 1970s is written, an important part of this story will be about the widening impact of econometric models on federal policy decisions. The wondrous computerized models—and the economists who run them—have become new rages on the Washington scene. These days it seems every spending and tax bill is played into mathematical simulations inside the computer. The model managers themselves are star witnesses before congressional committees whose members seek to define the future. And what these machines and their operators have to say has come to have a significant bearing on what Washington decides to do. . . .
>
> On more than one occasion the models have contradicted pronouncements by senior government figures. A small model at the Council of Economic Advisors, for instance, indicated that James Schlesinger, the Secretary of Energy, was talking through his hat last March when he predicted that disastrous consequences would flow from last winter's coal strike."

The examples go on and on of the recent use of large-scale econometric models in the debate over national economic policy. The most prominent among these large-scale models used in economic policy planning by government include DRI, Chase and Wharton modeling efforts.

Outside of the use of econometric models in general economic planning by government there has been a major surge of new computer modeling for purposes of energy policy planning and analysis. Some of the better known of these modeling efforts include the Department of Energy's (DOE) Project Independence Evaluation System (PIES), now known as Midterm Energy Forecasting System (MEFS), the Brookhaven Energy Reference System, the Stanford University PILOT modeling effort, the Baughman-Joskow Regional Electricity Model, and many others.

Greenberger, et al, in a recent book[2] have classified the types of models by five categories related to disciplines which are listed as (1) linear economics, (2) operation's research, (3) statistical economics, (4) urban and regional development, and (5) engineering. The nine specific methodologies listed within these disciplines include (1) input/output analysis, (2) linear programming, (3) two-person zero sum games, (4) probabilistic methods, (5) algebraic methods, (6) econometric modeling, (7) microanalysis, (8) land use analysis, and (9) systems dynamics. The disciplines of economics, operations research, statistics, urban and regional planning, and engineering are heavily involved in these quantitative approaches to modeling systems of the real world.

In energy policy analysis these models have been used most recently for projecting the world crude oil situation, the impact of U.S. Government conservation and production incentives on the U.S. economy, in producing annual outlook reports for the U.S. energy situation, and most heatedly in the national debate over natural gas pricing during the discussion of President Carter's National Energy Plan (NEP).

The above quotations simply serve to point out the fact that large-scale models are very prominent in government policy planning work and are being used extensively—large-scale computer models are part of the professional and political life of today.

The current dilemma of policymakers in the use of large-scale models is further exemplified by a recent interview with Secretary Schlesinger:[3]

Q: The original NEP would have saved how many barrels of imports?
A: We estimated 4.5 million b/d.
Q: So the level would be about 7 million b/d in 1985?
A: Right.
Q: And this year's legislation would save how much?
A: About 2.5 to 3 million b/d.
Q: So the level of imports in 1985 would be 9 or 10 million b/d?
A: Yes. Our estimate is in that range. Those models (which are used in forecasting) aren't worth very much.
Q: I'm glad to hear you say that—after the bill is passed you admit it, is that it?
A: I stated it rather forcefully from the first. Matter of fact, my troops had to force me to include these estimates from the PIES model. What we are trying to do is to change behavior, change behavioral reactions. Yet we get our estimates of the future from a model that draws on parameters of past behavior. It's just logically inconsistent.
Q: A model cannot predict the future—it can only estimate based on history?
A: Another thing wrong is that we should never have had only point estimates—we should have had a range. The volume of oil imports is contingent on such factors as nuclear power plant start-ups, the amount of natural gas produced or coal used, etc. I find it exceedingly difficult to predict how much natural gas we will produce in 1985. The resource base there is much more substantial than the oil base. How much really depends partly on whether intrastate producers sulk, believe in the "regulatory nightmare" myth, or whether they will just get cracking.

Q: Do you think the resource base on gas is still pretty large? Potentially undiscovered? You feel better now than a year ago?

A: Just look at the results. I'm not sure the numbers are different. I feel a little better about gas.

Q: You think the higher price of gas will encourage more exploration? More development?

A: We're estimating production of 2 trillion cubic feet more gas by 1985 in the lower 48 states under this legislation, compared with the status quo."

The Senate Energy Committee recently asked Dr. Herman T. Franssen, formerly with the Congressional Research Service, to look into the subject of forecasts of energy supply and demand, both national and international. Dr. Franssen's main conclusion is contained in the summary of his report.[4]

> The main reason for inaccuracies is that forecasts are influenced by the Zeitgeist—the spirit of the times. When the spirit is optimistic, the energy forecasts are optimistic. When all is gloom and doom, then everyone says we are running out of energy. Mathematical models, which are supposed to screen out such subjective influences, have been even less accurate than the subjective forecasts. As for the conflicting forecasts being made now with an eye towards 1990, one is no more likely to be accurate than another. Due to physical and human phenomena subject to surprise and frequent change, it is understandable that different projections of energy demand, supply, and prices can be believable at the same time. Finally, a lot depends on who hires the forecaster. The policymaker is no doubt aware that energy forecasters work for different clients, and that, for example, oil companies plan for future opportunities while governments have to plan for national supply security.

Seen from a slightly different perspective, Greenberger, et al, speak of the difficulty of modeling systems which include sociological behavior of people.[5]

> Theory as a basis for a formal model is of greatest value when it refers to a reference system that changes only slowly. Unfortunately, policy areas are among the most volatile fields of application for models. Economist Robert M. Solow of MIT puts it this way: "One advantage the physicist has over the economist is that the velocity of light has not changed over the past thousand years, while what was in the 1950s and 1960s a good wage and price equation is no longer so."
>
> Theories that change only slowly over time may not keep up with policy areas that change rapidly and consistently. Recognizing this shortcoming in theory a modeler may personally adjust the projections of his model to reflect an impending strike, war, bankruptcy or other events and trends not covered by the theory. In a model of the national economy "add factors" may be applied to the modeler's projections to

incorporate the modeler's judgment and intuition about economic trends. The practice is widespread. Many well-known econometric models use some form of judgmental adjustment. Louis Mumford has observed recently that computer models can acquire a God-like authority and the results can be taken as gospel. A model masquerading as an oracle may be nothing more than an advocate in technological guise.

However, policymakers must use large-scale models despite a scepticism of their predictive reliability, because they have no alternative way to handle these complexities. Secretary Schlesinger's interview exhibits this dilemma on the part of policymakers.

The major task that lies before us is to improve the usefulness of models and the judgments of professionals involved in public policy analysis. The central issue is the procedures by which the reliability of large-scale models, especially those used in public policy work, can be established and made transparent—to distinguish between the influence of professional, subjective judgments and the influence of information that is reproducible by others. The Texas National Energy Modeling Project (TNEMP) has made some contribution to the goal of increased model credibility by transferring and operating MEFS.

PROJECT PURPOSE

The first purpose of TNEMP is to provide an independent evaluation of the Energy Information Administration's (EIA) MEFS. The evaluation will provide guidance to users of MEFS concerning the level of confidence one may have in the results of the models for government energy policy analysis purposes. The evaluation is critical, in the best sense of the term as used in scientific work, but also suggests improvements in the model structure and in procedures used by the EIA for increasing model credibility.

The second purpose of TNEMP is to provide recommendations to the Texas Energy Advisory Council (TEAC) concerning the maintenance of a national modeling system by the Council to evaluate Texas impacts within a consistent national modeling framework. As a result of the exercise, we have first-hand experience with MEFS, as well as Data Resources, Inc.'s (DRI) macro and energy models. We are able to compare these models, their structure and results with various Texas models at the TEAC, the University of Houston, and The University of Texas for purposes of assessing their relative usefulness for energy policy analyses and possible joint use or integration. TNEMP participants have experience in still other modeling efforts, as well as experience in model evaluation, institutional arrangements for housing models, and procedures for making the best use of resources to achieve successful model development, use, and credibility. TNEMP is well suited for the second purpose.

The evaluation of Texas impacts of national energy policy decisions must necessarily be done within a consistent national framework, since a major portion of the nation's energy production and processing, as well as corporate management of these for the nation as a whole, is in Texas. A significant fraction of the nation's energy is also consumed in Texas, especially natural gas. Texas-based energy analyses must, therefore, be national to properly reflect Texas' interests and its role in the nation's energy future. In order to be of use in the national energy policy process,Texas-based analyses must be centered in a credible national modeling and analysis framework.

A third purpose of TNEMP is to raise the level of attention concerning the current uses, practices (potential as well as current abuses), and potential dangers of modeling for developing public policy and clarifying important issues to the citizens of a democratic society. The increased reliance of policymakers, high-level advisors, and the voting public on expert opinion, based in part on large-scale data bases and models, in an era of complex problems and policy prescriptions, requires that policymakers and high-level experts gain a better understanding of current modeling reliability and practices. TNEMP is intended to help achieve this goal.

Two additional factors are important side effects of the TNEMP exercise. First, we have advanced by some degree the art of model evaluation by achieving, among other things, the transfer and operation of a large and complex model. We hope others will benefit from both our successes and our mistakes; for model evaluation is far from being a well-developed discipline. Second, we hope that some insight has been gained into what kind of institutional setting is appropriate for third-party model evaluation efforts. The current institutional arrangements for the building, operation, and maintenance of large models are critically lacking in efforts aimed at validation of relations, assumptions, data verification, and model documentation. The quest is for the development of model evaluation institutions and/or incentives to bring about a timely devotion of model developers to validate, verify, and to provide clarity and workability as model attributes. Third party evaluation can increase the probability of a model being accepted. It can enhance the transfer of science, technology, and statistical data to the policy decision process.

ORGANIZATIONAL STRUCTURE FOR THE STUDY

The organizational structure for the study consisted of four primary groups. First, the TEAC Executive Director provided project direction and the staff provided coordination, materials, and other support. Second, the *National Advisory Board* provided advice on procedures, suggestions on methodology, evaluation of the integrity of the evaluation process, and general recommendations for a Texas national modeling capability. Third, the *Analysis Team* provided an evaluation of MEFS, made recommendations for improvements and alternatives, and made specific recommendations for Texas maintenance of a national modeling capability reflecting the impact on or by Texas. Fourth, the *Supporting Institutions* provided support by endorsing the objectives of the study, making review and comment, making data and facilities available, and funding the study. *Figure 1* illustrates the working relationships of the four groups.

Meetings were held periodically for the purpose of reporting progress of various studies to the National Advisory Board, other members of the Analysis Team, and Supporting Institutions. This format provided opportunity to refine project objectives, define evaluation criteria, share reference material, identify weaknesses in the project study design, interact project participants with DOE personnel, and provide overall guidance for the Project Director and individual principal investigators.

In order to provide credibility of the project objectives and procedures, the National Advisory Board was asked to write an evaluation of the study to be published with the final report; their statement makes up Chapter 4. In order to encourage maximum intergovernmental cooperation, EIA was offered the opportunity to comment on the project with the assurance that the remarks would also be published with the final report. EIA comments are included in Chapter 5. To achieve the objective of familiarizing key decision makers with the important findings of the project, a meeting was arranged between members of the National Advisory Board, officials of EIA, and the Lieutenant Governor of Texas, who serves as Chairman of TEAC.

EVALUATION CRITERIA

The reliability of information from the EIA large-scale energy model known as MEFS, currently being used for forecasting, policy evaluation and analysis, is our primary concern in this study. The determination of reliability is a difficult task; much has been written recently on the topic but not much is well defined and clarified at this time. We have made certain interpretations concerning the process and measures to be used in assessing reliability and usefulness. In the usual understanding of reliability in scientific areas, one would expect modeling results to be unambiguous, reproducible, and transferable. Such characteristics need to take on special definitions however, since we are dealing with a system involving the behavior of people (the economy with emphasis on the energy sector) and are making applications of models to describe alternative worlds, some of which will never exist for testing purposes. We also need to be specific with respect to the user of the information, and in our case this will involve both other modelers and professionals as well as policy decision makers and the public at large. It is also essential to distinguish uses of the models between

Figure 1.

Texas National Energy Modeling Project Organizational Structure

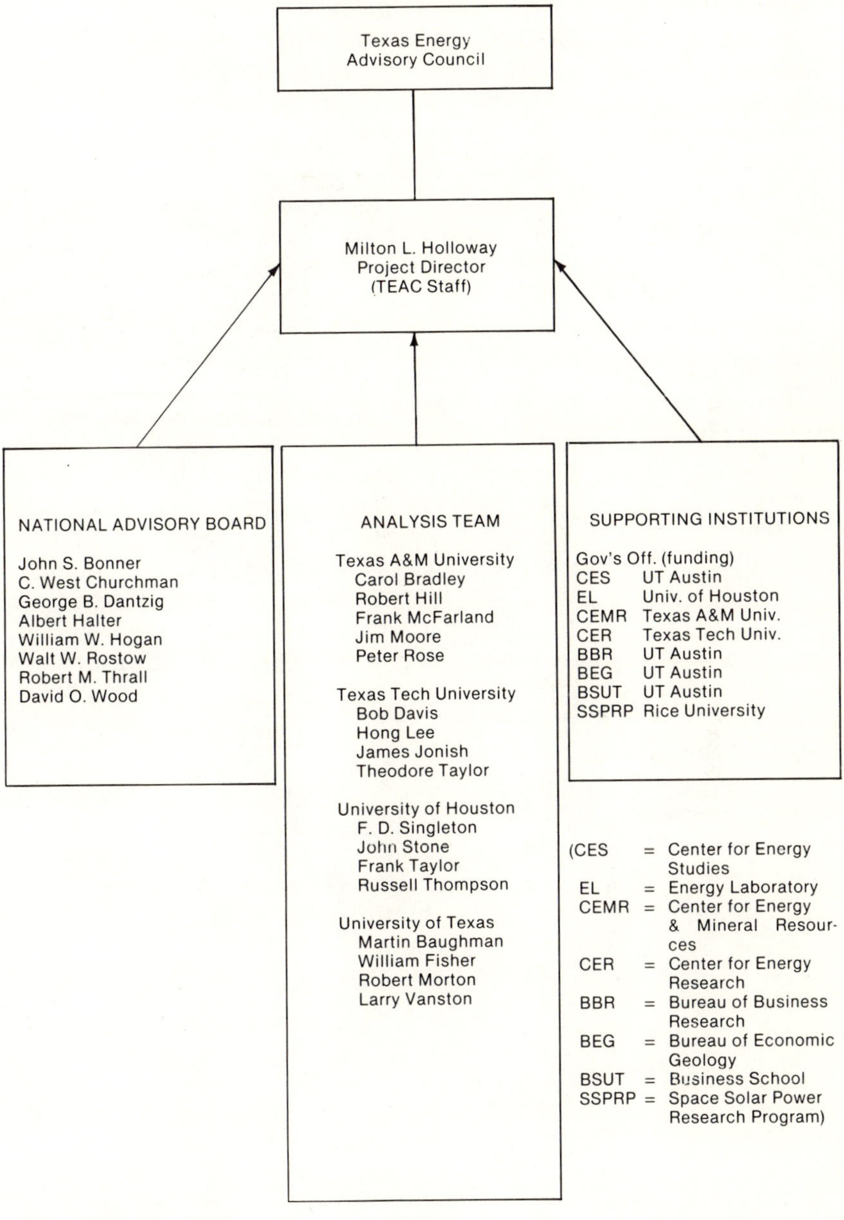

raising issues versus resolving issues. Models used for raising issues may rely more on hypotheses for their formulation, whereas models used and designed for resolving issues will necessarily have to be based on accepted theory and/or laws.

MEFS of the EIA is used primarily for the purpose of resolving energy issues, and their use is for policymakers and the public at large. For that reason the evaluation criteria we have selected for TNEMP take on specific definitions.

There is currently no consensus on either the appropriate set of criteria for evaluating models or the definitions of commonly used terms in model assessment literature and discussions. Various discussions of model assessment criteria may be found in recent publications by the MIT Model Assessment Group, Saul Gass, Judith Selvidge, A. N. Halter, and others.[6]

However, four common themes run through the discussions and form the basis for our evaluation criteria in TNEMP. The criteria are workability, clarity, validity (coherence), and verifiability (correspondence).

The four themes commonly discussed in connection with model assessments are interrelated. First, if the model is to serve any useful function other than the satisfaction of the intellectual curiosity of the modeler, then it must serve some practical function—identify issues, suggest a resolution of issues or in general it must produce results that relate to the practical problems of the user. We have interpreted this theme as our criteria of workability. In the context of the energy model which we are evaluating, we interpreted workability to mean that the model was capable of providing information regarding such things as the inflationary impacts of decontrol policy, the impacts of conservation and decontrol on the import levels of crude oil, the effect of the combination of energy policies on international trade and the value of the dollar, the impacts of deregulation, coal conversion and other such policies on economic growth, and perhaps the regional distribution of economic growth, as a result of national policy. So in the context of our model evaluation effort the model is workable if it raises important policy issues in these areas, points toward the resolution of important issues, or provides explicit information pertinent to an issue.

Second, it is implicit that a model cannot have good workability characteristics if it is not clear. Our interpretation is that clarity must be defined in terms understandable to other modelers (who, for example, may provide a function of advisor or interpreter to laymen and/or policymakers) and in terms understandable to laymen and/or policymakers, who are the users. For our model evaluation effort this means that the model's behavior and its results must be translatable into various supply and demand representations of the economy, understandable to economists, and intuitively comprehensible by laymen who want to know the income, inflation, tax, energy costs, employment, and economic growth implications of the policies being analyzed. This understanding is embedded in our use of the criteria of clarity.

Third, if a model is to have workability and clarity, it must be unambiguous. The model must be coherent in that it must behave in a manner consistent with the underlying logic or theory upon which it is built. Furthermore, the model must do what the modeler says it will do as per its design. One should expect greater difficulty in achieving acceptance of the model by the disciplines if the basic theories of relationships are violated or if one is attempting to develop new or modified theory in the context of a model designed to resolve policy issues. To develop a model representing a market economy in the absence of accepted economic theory, is to challenge the basis for the economics profession; use of the model will, therefore, have a much different effect than a model based on well accepted economic theory. Coherence is at least partially dependent upon the understanding of the disciplines best represented by the current body of theory. In terms of our model evaluation (since MEFS is explained by EIA to be a model of the U.S. energy and macro-economics) we should be able to verify that increased prices for oil and gas simultaneously bring on increased supplies, reduced demands for oil and gas, and increased use of alternative energy sources. In some sense the model must behave in a manner consistent with the economic theory of markets in the context of the U.S. domestic economy and if they do not the theories or model require modification. This theme is consistent with our criteria of validity or coherence.

Fourth, there is the question of the relationship of the model to "reality." It is tempting to state the fourth criterion (verifiability) in such phrases as "the model ought to represent reality." Such a criterion works reasonably well for model airplanes or planetarium models of the solar system, because we can observe both the reality (real airplanes or real planets) and the model results or consequences and identify some correspondences. But the reality of an "energy system" and "the economy" cannot be observed in such a simple manner. Instead, there are data banks, usually but not always put together by the disciplines, that are at best major abstractions and aggregations. It is reasonable, however, to expect that the model results should agree reasonably well with the accepted data; but if it does not, then some satisfactory explanation should be forthcoming; e.g., the data are incorrect or incomplete, or the models require modification. The model consequences or results must correspond with reality in this sense. This theme is consistent with our criteria of verifiability or correspondence.[7]

In TNEMP we have considered this set of four criteria from the modeler's, the model assessor's, and final user's point of view. The implication is that the modeler should use such criteria and further, that a responsible modeler would do so. The criteria are also useful from the viewpoint of the model assessor. MEFS is judged by our group on its workability characteristics with primarily a user's point of view of important practical problems; to a lesser extent the models were evaluated on the basis of their ability to raise important issues or to help resolve important issues from an analyst's, interpreter's, or advisor's point of view. We have paid particular attention to the *apparent* intended uses of the model (as evidenced by its historical applications). Validity was judged in our project from the analyst's and final user's points of view. Extensive comparisons were made with the current body of economic theory, since the model attempts to represent the market economy. We also paid attention to MEFS's physical relationships of geology and engineering, which also must be coherent with accepted theory in those fields. Validity work was not documented by the modelers in this case, and was judged by model assessors by comparing the model performance based on our own operation of the model with documented expectations of the model builders and with logical deductions based on the theoretical underpinnings of the modeling system. Verification work was not documented by the modelers either. Work on verification by the model assessors was not extensive since it was impossible to operate the models comprehensively to obtain results for time periods when data is available, exclusive of that used to estimate model parameters.

FOOTNOTES CHAPTER 1

[1] Cameron, Juan, "The Economic Modelers Vie for Washington's Ear", *Fortune,* November 20, 1978, p. 102.

[2] Greenberger, Martin, Matthew A. Crenson and Brian L. Crissey, *Models in the Policy Process: Public Decision Making in the Computer Era,* Russell Sage Foundation, New York, 1976.

[3] Bachman, W. A., "What the Industry Can Expect Now From the Department of Energy", *The Oil and Gas Journal,* November 13, 1978, p. 156.

[4] Franssen, Dr. Herman T., "Energy: An Uncertain Future," Committee Print 95-157, December, 1978.

[5] *Op cit,* p. 67.

[6] See for example, "Evaluation of Complex Models" (Gass, Saul, 1977), "A Procedure for the Evaluation of Complex Models" (Gass, Saul, 1977), "Strengths and Limitations of Models from a User's Viewpoint: Modeling the Modeler's Model" (Halter, A. N., 1977), "Independent Assessment of Energy Policy Models: Two Case Studies" (The MIT Model Assessment Group, 1978), and "Panel Discussion—Management Audit of Quantitative Models" (Selvidge, Judith, 1978).

[7] I am indebted to Glenn L. Johnson for pointing out that I had mixed the definitions of validity and verifiability in the final draft prior to printing of the manuscript. Johnson also points out that the meanings of the two terms are inverted in the systems science literature as compared to the philosophy of science literature. The main concern here is not to resolve this difference in the literature but to be consistent throughout this writing in the use of the terms. I hope I have done so.

CHAPTER 2

ANALYSIS TEAM CONCLUSIONS AND RECOMMENDATIONS

by Milton L. Holloway

INTRODUCTION

The Texas National Energy Modeling Project (TNEMP) commissioned eleven studies which fall into four groups: (1) computer operations of the Mid-Term Energy Forecasting System (MEFS)—study 1; (2) supply modeling—studies 2, 3, 4, 5, and 6; (3) processing and transportation modeling—studies 7, 8, and 9; and (4) demand modeling with macroeconomic interface modeling—studies 10 and 11. This set of studies necessarily deals with physical relationships in engineering and geology, industry and consumer economic behavior, and the economics of conditional market equilibrium, since this is the domain of MEFS.

Since MEFS is a large and complex modeling system developed over several years, with many man years of effort and with the expenditure of several million dollars, it was not considered practical to evaluate every aspect of the model and the data. Therefore, the *initial* study design focused on the evaluation of (1) the crude oil, natural gas, and coal satellite models with hands-on operation on the computer; (2) the transportation, electric power, and refinery (and synthetics) sector representations in the integrating linear programming model; (3) the macro/microeconomic interface in all its aspects and the specific formulation of the demand models with hands-on operation of the Data Resources, Inc. (DRI)/MEFS interface on the computer; and (4) the integrating model with hands-on operation. The 1977 National Energy Plan (NEP) version of MEFS was transferred to Texas for purposes of the evaluation. Figure 2 shows the portions of the modeling system evaluated in this study. The NEP version of MEFS was "brought up" on the Texas A&M University Amdahl computer, and the principal investigator in charge of the system was to provide computer and operating support for other members of the team. Evaluation of other supply data and related exogenous calculations, including uranium for nuclear power production, solar and other new technology contributions on the supply side, data and programs for estimating conservation impacts on the demand side, and data and calculations specifying non-crude oil import levels, were to be ignored in the evaluation.

A mid-course correction in the study implementation set priorities for the hands-on operation on the computer to emphasize: first, the testing of the oil and gas supply model behavior with impacts on the integrating model solutions; second, the operation of the coal supply model with impacts on the integrating model solutions; and third, the operation of the system to test macro/micro model interface results and the impacts of errors or varia-

Figure 2.

Simplified Overview of Midterm Energy Forecasting System Components*

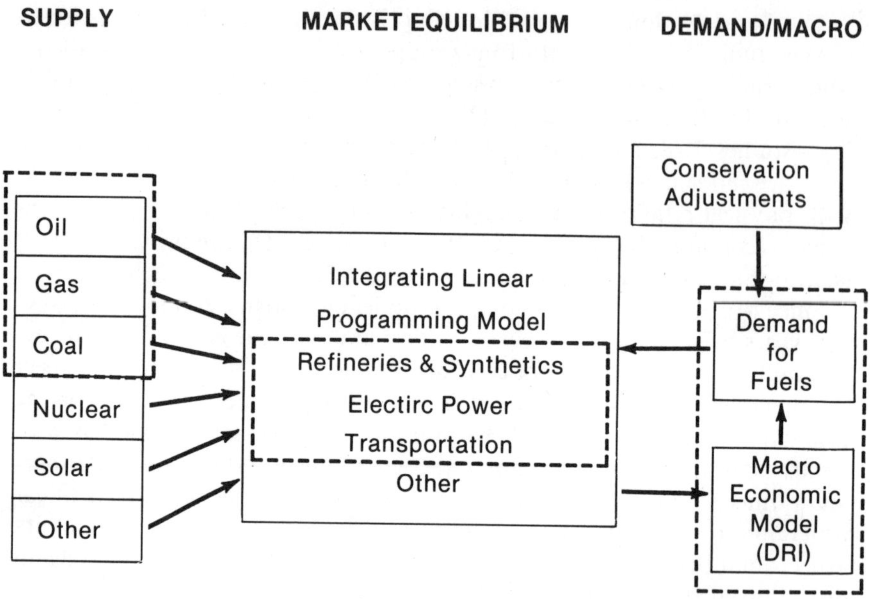

*Circled components indicate portions of MEFS evaluated in this study. The oil, gas, and coal supply models were operated on the computer using selected parameter variations for sensitivity analyses; the demand model was mechanically operated on the computer but sensitivity analyses were done by ad hoc adjustments to demand model parameters; existing results of the DRI model were used to test the sensitivity of the macro economy impacts of energy market shocks; and the integrating model was operated many times to incorporate pair-wise comparisons of market equilibrium solutions from changes in the demand, oil, gas and coal supply models.

tion in demand model parameter estimates. During the October 1978 National Advisory Board meeting, including discussions with Energy Information Administration (EIA) personnel, it was determined that the NEP version of MEFS could not be verified by either EIA or TNEMP, nor could TNEMP operate it intelligently, due to poor and non-existent documentation. Therefore, it was agreed that TNEMP would focus instead on the 1977 Administrator's Annual Report version of the model. This version was accordingly transferred and operated on the Texas A&M computer for TNEMP evaluation purposes.

MAJOR CONCLUSIONS CONCERNING MEFS

This section summarizes the major conclusions about the extent to which MEFS and its major submodels satisfy the criteria for evaluation. The conclusions are, to the extent possible, kept within the context of the state of the art for large-scale models in general, and energy policy models in particular. The statement is also kept within the context of the modeling system's purposes as indicated by how it has been used, to the extent we were able to ascertain such uses. (See the Appendix to Part One for a listing of major uses of MEFS). This section deals specifically with conclusions concerning the reliability of major model outputs used in energy policy applications. To the extent possible, conclusions are drawn concerning the extent to which major model outputs are a function of exogenous-to-endogenous variables, and thus the explanatory power of the models.

MEFS Operations (Study 1)

Study 1 deals primarily with the questions of clarity, workability, and transferability of MEFS. This system is a complex combination of a number of formal models, data bases, parameter specifications, and related software packages for operation and report writing. The need for the ability to transfer and operate large-scale models used in public policy analysis is crucial to the overall establishment of credibility and perhaps a major requirement for completing tests of validity, verifiability, workability, and clarity.

The basic questions addressed in this study include (1) whether MEFS can be successfully transferred and operated, (2) whether documentation is adequate for clarity, and (3) whether various simulations can be performed for purposes of validity, verification, and workability tests. The study further provided for operation of successfully transferred models as a service function to the other principal investigators for individual study testing and operation.

The set of software systems installed and satisfactorily tested include:

1. Midterm Energy Market Model (MEMM)
2. Regional Demand Forecasting System (RDFOR)

3. RDFOR Preprocessors
   a) Price Preprocessor
   b) Marco Preprocessor
   c) Parameter Preprocessor
   d) Initial Prices Preprocessor
   e) Conservation Shifts Preprocessor
4. Midrange Coal Supply Model
   a) does not include the entire National Coal Model (NCM)
   b) two University of Houston Fortran Modules have replaced the need for the Gamma Matrix Generator and Report Writer
5. Midterm Oil and Gas Supply Model System
6. Standard Integrating Table Preprocessors
   a) Oil Preprocessor
   b) Gas Preprocessor.

The various systems were installed and results checked against Department of Energy (DOE)/EIA results from the same systems operated on the DOE computer in Washington, D.C. Minor differences in the DOE check results and those produced at Texas A&M have been attributed to the following:

1. Version 1 of MAGEN used at Texas A&M versus Version 2 of MAGEN used at DOE
2. MPSX/360 used at Texas A&M versus MPS3 used at DOE
3. Different starting basis
4. Possible different alternate solutions
5. Replacement modules used in the Coal Supply Model.

The check results were exact except for the integrating system and the coal supply model. The base coal supply file produced by the modified coal supply model was input via the coal preprocessor and the final "cookie" (final report writer results) results were very close to the base "cookie" produced with DOE's AAR77 standard coal supply table. The differences in the integrating model results are in general quite acceptable. Some question has arisen in the area of non-associated gas results. DOE personnel suggested the possibility of alternate solutions when this was discussed with them. Some differences have been found in the output of Version 1 of MAGEN, but this is not of major concern.

The major difficulty of implementation has been the incompatibility of the MVS operating system used at DOE versus the SVS system used at Texas A&M University. Many job streams had to be broken down into multiple jobs in order to prevent various queues from overflowing. Secondary difficulties were different physical intermediate I/0 devices, lack of DASD allocation, required in-or-out specifications on the label parameter, and the inability to rewind DD * data files.

Many scenarios were run at Texas A&M in order to test the sensitivity of the overall model to major changes in selected parameters. The demand model was analyzed by initially varying the elasticities produced from RDFOR. The consumption levels for fuels changed very little. The implication is that price paths which drive the model and are input to the model (predetermined) must be changed to significantly affect demand. This result was not clear at the outset. Shifts in the gas supply functions were combined with changes in the elasticity estimates in order to trace out a portion of the gas demand curves and thus determine the importance of variations in the elasticity estimates in the demand model. These tests were done successfully.

Variations were performed on key parameters of the oil and gas supply models including (1) finding rates, (2) discount rates, (3) price paths, and (4) rig life. The variations were completed satisfactorily to allow analysis of the importance of these parameters on oil and gas production response behavior of the models.

Variations were also done on the "mine life" parameter of the coal supply model. These variations were completed satisfactorily allowing analysis of coal production response behavior of the model.

The oil, gas, coal, and demand model variations were interfaced pairwise with the integrating model in order to test the sensitivity of the overall market equilibrium solutions to the parameter variations.

One case (low finding rate for oil and gas) produced an unbounded solution problem in the integrating model. This occurred when attempting to insert the oil and gas supply curves corresponding to the 50% lower finding rates. Without the aid of DOE, this problem would have required an enormous amount of time to solve; one must thoroughly understand the workings of the model if such problems are to be solved. This was a new problem never encountered by DOE, implying the need for more verification tests by DOE or others using the model.

MEFS is definitely not of the "push button" variety and requires great attention to detail in the preparation of each scenario. However, the overall operation is quite straightforward and should be mastered by most reasonably proficient system analysts.

MEFS (PIES) was not designed with transfer in mind, although past and current employees of DOE (FEA) have said that in principle it was always thought that it could be transferred. The partial transfer accomplished in TNEMP is therefore a major achievement very desirable for DOE and TEAC as well as for the general enhancement of modeling for policy analysis. The primary lesson to be learned from the experience is that the current state of the art in large-scale models lacks badly for clarity sufficient to accomplish easy transfer; major personal communication and large resource commitments are required to achieve effective transfer.

Supply Studies (Studies 2, 3, 4, 5, and 6)

Three factors are important to supply modeling and projection work, especially in the case of the oil and gas resources. First, the estimated size of the resource base as well as that portion which can be economically recovered (economic reserves), given current prices and technology must be known. Second, the physical relationship between drilling on the intensive and extensive margins and added reserves (finding rate) must be understood and estimated. Third, the behavior of the industry must be understood concerning investment, drilling, and production decisions in response to expected prices of oil and gas and changing expected costs of drilling and production. The same basic factors are required for modeling coal production; i.e., one must know the estimated reserves and economic reserves, physical recovery rates and economic behavior of the industry.

Study 2 deals primarily with the questions of validation and verification in the EIA oil and gas model. The evaluation provides the basis for showing the importance of estimates of economic reserves and finding rates on future production levels. The methodology used by EIA in projecting the finding rate for oil and gas, as the number of wells drilled increase and the remaining resources decrease, was evaluated to determine the adequacy of the time path specifications used in the projection work.

The basic questions addressed in this study are whether or not empirical estimates of economic reserves and findings rates may be of such large variation as to cause significant variations in output of the models regarding possible future production levels.

Two major conclusions summarize the study. First, the resource volume numbers, even given the uncertainty of undiscovered resource estimates, should not be a significant constraint in short- and intermediate-term projections (through 1985); it becomes critical, however, in long-range projections (1990 and longer). A reasonable variation in the resource numbers used by MEFS would not be expected, therefore, to *directly* affect short- to intermediate-term projections. Second, historical data exhibit a distinct post-1970 exploration epoch, for which the finding rate has been relatively stable, and thus reliable for short-term (through 1985) projections. Analysis of regional shifts in the drilling mix show that these have very little effect on calculated finding rates. It is concluded that forward projections of reserve additions should utilize a finding rate in the range of 17-18 barrels of oil and gas equivalent per foot rather than the 21.5 boe per foot average implied in the NEP projections or the higher 24.7 boe per foot average utilized in EIA's Projection Series C through 1990.

The DOE calculation of finding rate, applying a continuous decline function to the past twenty years of historical data constrained by U.S. Geological Survey (USGS) reserve estimates, yields projected economic reserve additions 20-40% higher than recent history of finding rates would seem to warrant, and thereby underestimating the drilling effort necessary to support given production levels. The finding rate specification in MEFS is intuitively sound but is questionable empirically and lacks a valid basis for specification. Further research is needed to better identify the factors underlying observed finding rates for use in modeling. Study 5, which follows, summarizes MEFS behavior under varying finding rate specifications.

Study 3 deals primarily with questions of validity and verification in the EIA oil and gas supply models. By studying the investment decision framework of the EIA models and the current theory of investment behavior, the study is able to make comparisons for the validity test. By studying empirically how the industry behaves the study provides direct and specific evidence of the oil and gas industry decision making process.

The basic questions addressed in this study are (1) whether the investment process depicted in the oil and gas supply models adequately reflect current investment theory, (2) whether industry specific empirical information supports or perhaps adds to understanding of the investment process, (3) whether empirical information from survey work can add reliability by incorporating empirical information into models of oil and gas supply, and (4) whether empirical evidence supports the idea that cash flow, in addition to marginal prices and other factors, influences investment decisions of oil and gas firms.

Four major conclusions characterize the study. Generally, the study indicates that a particularly serious problem is the apparent inconsistency of the investment model with capital budgeting theory and recent developments in risk measurement. First, the net present value (NPV) model was designed for the evaluation of a single investment project at the margin, not for determining regional drilling activity from average price and cost estimates as used in the oil and gas supply models. The regional approach averages out differences among individual firms and capital projects. This appears to introduce a bias in drilling activity at both high and low marginal prices. Alternative specifications of the investment process should be tested. Investment behavior models appearing in the recent econometric literature have generally relied upon both the demand for capital goods and the marginal capital supply function.

Second, the rate of discount assumed to motivate investment in the model appears to be seriously underestimated. A real discount rate of only 8% probably substantially underestimates the risk inherent in petroleum operations. A survey of petroleum companies suggests that a real discount rate of at least 15% after taxes may be more appropriate. The model was operated on the computer using a variety of discount rates to examine the consequences of changes in the cost of capital on the volume of drilling activity undertaken, and therefore on production. The model results for a change in the discount rate from 8-15% show an average (arc) elasticity (percent change in 1985 production from a 1% change in the discount rate) of -.578 for natural gas and -1.825 for crude oil at $1.75 and $12 prices respectively. The elasticities vary greatly over price ranges specified, ranging from -3.286 for gas at $1.00 per mcf to -.166 for gas at $3.00 per mcf and from -3.286 for oil at $10 per barrel to -.497 for oil at $16 per barrel. Further sensitivity results are summarized under Study 5 below.

Third, probably the most serious shortcoming of the model from the standpoint of financial theory, is its failure to incorporate risk. To the extent that investment projects adopted by the industry add significantly to the total risk complexion of the firm, the model would tend to overestimate the volume of petroleum investment spending.

Fourth, survey results from a sample of independent drilling companies suggest it is inappropriate to assume that all petroleum firms use the NPV approach in making their investment decisions. A variety of approaches are used in practice, and there is evidence that petroleum firms make use of NPV, pay-back, internal rate of return, and accounting-based methods in combination.

The survey results also suggest the critical importance of marginal prices on new output in determining petroleum investment. Net cash flow directly affects the industry's marginal cost of capital which influences its investment. Revenue generated by current investment directly affects industry sources of financing through an impact on the availability of internally generated funds. This circular relationship between cash flows, cost of capital, and investment suggests the necessity for a multi-equation framework within the model to more realistically reflect petroleum's capital-budgeting process.

The industry's increasing use of debt as shown in recent published data has generated the potential for greater volatility in returns to stockholders. This factor would suggest that either the discount rate applied to expected cash streams will have to be increased or the variability of estimated cash flows (risk) included formally in the model if it is to more accurately mirror petroleum investment decisions.

Study 4 is primarily concerned with the question of validity in the EIA oil and gas supply model. By studying the behavior of the industry and examining secondary historical data on stripper well production, the project determines whether special consideration should be given this class of producers in models which attempt to project oil and gas supply response to price changes in the current complex pricing structure maintained by the federal government.

The basic questions addressed in this study are (1) whether the peculiar behavior of stripper well owners warrants treating this class of producers uniquely in oil and gas supply models, (2) whether the growing population of stripper wells (currently 73% of U.S. total oil wells constituting 14% of U.S. total oil production) is likely to make this consideration more important over time, and (3) whether empirical data can be obtained for modeling this sector of oil and gas production.

The study results can be summarized in one major conclusion. The oil industry in MEFS is treated as a monolith. The stripper well segment is aggregated into the rest of the industry. The critical abandonment-produce decision that is continually faced by the marginal producer is assumed away when the exploration drilling decision is made in the model. In short, MEFS does not appear to model the economic characteristics of the stripper well industry and this part of the industry is of growing importance. Further study and data gathering will have to be done to incorporate this sector into oil supply models.

Study 5 deals with questions of validity, verification, workability, and clarity in the EIA oil and gas supply models. By having hands-on operation of the models, the studies are concerned with the actual behavior of the models, testing the sensitivity of major output variables to endogenous and exogenous variables.

By performing such tests and by examining the actual equation structure of the models the study determines the extent to which the specification conforms to economic theory. By having hands-on experience with the models the analysts were in a good position to assess the criteria of clarity. The study was also able to press toward understanding the requirement for verification.

The basic questions addressed in this study are (1) whether documentation and other descriptive materials are adequate for a user to operate and understand the models (clarity), (2) whether basic functional relationships can be synthesized for comparison with economic theory (validity), (3) whether the models answer the questions of production levels, drilling activity, and reserve additions needed by the user (workability), and (4) whether it is possible to provide tests of verification on these models.

Several highlights stand out from the evaluation. First, regarding validity, the model represents an attempt to be consistent with the theory of economic allocation of (drilling and production) resources, subject to declining finding rates and increased costs of drilling. However, the fundamental joint product relationship between crude oil and natural gas supplies has been totally ignored. Ignoring this joint product relationship significantly undermines the validity of the oil/gas supply models because drilling capacity in reality may be used in drilling for oil or for gas alternately, and further, some combination of oil and gas is generally found in every discovery. The result is to significantly bias the natural gas supply estimates downward and the crude oil supply estimates upward. This weakness raises a major question as to the validity of the model.

Second, use of a statistical finding rate estimate constrained by the USGS's estimate of reserve significantly modifies the projections of finding rates from that of an unconstrained statistical estimate. There is no consistency between the economics of the model itself and the implied economics embedded in the USGS economic reserve estimates. This constraint means the economics of reserve additions are totally confused by the reserve estimating procedures of the USGS. Institutional jurisdictions and responsibilities (reserve estimates by the USGS) have seemingly forced the imposition of a constraint, which may have nothing at all to do with the economics of additional oil and gas supplies. But this constraint has clearly modified the finding rate projections used in the oil and gas supply models, the supply projections made by the models, and the economic results obtained from the integrating model.

The finding rate specification procedure in the model is conceptually nice but needs empirical content. That is, more empirical work is needed here to explain the finding rate behavior. No variations on this parameter are commonly shown with the EIA model results even though this factor is of obvious and great importance in production projections. For example, the behavior of the gas model, when operated on the computer by analysts in Study 5, shows very little production response to price changes near the current market price for the base case (parameter specifications by EIA for the 1977 AAR version of the model). Production responses in the gas model to *marginal* price changes show that for the base case, production is very insensitive to price changes in the \$1.75-\$3.00 range, decreasing drastically from that in the \$1.00-\$1.75 range. For the lower finding rate case (50% decrease), however, production response to price changes is shown to be large even at high prices. Thus, the finding rate specification (or USGS reserve estimates by which the finding rate is constrained) greatly affects the price responsiveness behavior of the model.

The specification of finding rate also has a large direct impact on expected *production* levels at a given price set over the entire price range tested. For example, the elasticity of production with respect to finding rate for gas ranges from 1.131 for high finding rate/\$3.00 gas to 3.000 for low finding rate/\$1.00 gas in 1985.

These findings have significant policy implications since (1) there is a considerable question as to the proper finding rate specification (see Study 2) with strong empirical indication that the EIA rate is too high by 20-40%, and (2) the model has been used in national energy debates to show that gas production response to price is very small at prices above $1.75. This model result needs further attention before the model is used for other policy analyses. The model suffers on the verification test.

Third, on the matter of clarity, the basic conceptual framework, once discovered, is reasonably straightforward. In implementation, however, the model is massive, complex, and convoluted. This situation, combined with incomplete (and occasionally erroneous) documentation, renders comprehension of the model unnecessarily difficult. Simply stated, the documentation would fail any reasonable test of clarity by even the most understanding modeler.

Fourth, with respect to workability from the modeler's point of view, MEFS's oil and gas supply models are cumbersome and difficult to operate, their efficiency and flexibility suffering greatly from the model's patchwork origins from the National Petroleum Council model. From a user's point of view, the limited purpose model would be strained to answer more detailed questions about petroleum industry financial behavior or the impacts of complex regulatory measures now placed upon the industry.

Study 6 deals with questions of validity, verification, workability, and clarity in the EIA coal supply model. By having hands-on operation of the model on the computer, the study is concerned with the actual behavior of the models, testing the sensitivity of major output variables to endogenous and exogenous variables.

By performing such tests and by examining the actual equation structure of the models, the study determines the extent to which the specification conforms to economic theory. By having hands-on experience with the models, the analysts were in a good position to assess the criteria of clarity. The study was also able to press toward understanding the requirements for verification.

The basic questions addressed in this study are (1) whether documentation and other descriptive materials are adequate for a user to operate and understand the models (clarity), (2) whether basic functional relationships can be synthesized for comparison with economic theory (validity), (3) whether the models answer the questions of production levels and regional distribution of production from alternative demand scenarios, and (4) whether it is possible to provide tests of specification on the model.

Several highlights stand out from the evaluation of MEFS's coal supply model. First, the coal supply model is a valid representation of coal supplies in the intermediate term for which it was designed. In the long term, the model has no provision for retiring "new" mines to account for the dynamics of coal depletion.

Second, concerning verification of the model, it adjusts faster than is possible in the industry because of lag times between building decisions and production start-up not accounted for in the model. The specification of one single lifetime for all mines in all regions needs verification with industry representatives. Mine life should be allowed to vary for different types of mines and different regions.

Operation of the model on the computer verified prior results by EIA. Variations in mine life assumptions indicated an apparently significant change in the supply functions for 1985, but the overall effects on market clearing prices from the integrating model solution were small.

Third, documentation of the coal supply model is reasonably adequate since the basic ideas have remained unchanged from the inception of MEFS (initially PIES). Improved clarity has been achieved at the University of Houston by eliminating one of the computer languages. From a user's or layman's point of view the model concept is easily understandable.

Fourth, workability of the coal supply model is adequate for its intended purpose of defining regional supply curves for MEFS's integrating model. The model will answer basic questions of relative regional coal production at various prices, provided the transportation, electric utility, and other coal-using modules in MEFS are correct.

Process and Transportation Studies (Studies 7, 8 and 9)

Three sectors are important in transforming basic energy into usable energy forms and products for consumers, and in transporting major energy raw material among regions of the country. This set of studies examines the MEFS representation of refineries (and synthetics), electric power generation, and transportation of basic forms of energy including coal, crude oil, refinery products, and natural gas.

Study 7 deals with the questions of validity, verification, workability, and clarity in MEFS's representation of the refinery sector and synthetic fuels production, which is likely to be of importance as new technology develops. The representation of refinery sector operation, including the change in the mix of products in response to changes in the relative price of products and cost of various crude types, is examined.

The basic questions addressed in the study were (1) whether the model representation of the refinery sector corresponds to economic theory relevant to refinery operations (validity), (2) whether the physical relationships for converting crude oil into a range of products are maintained for a range of relative product prices (verification), (3) whether further development is needed to represent the synthetic fuels sector, and (4) whether the sector representations are capable of answering relevant questions within the context of the overall model (workability). The study amounted to an examination of the refinery sector representation in MEFS's integrating model including some computer runs to verify the behavior of the model.

Several conclusions summarize the study. The simple structure results in the total loss of technical structural delineations characteristic of the refining of petroleum products from various grades of crude oil, thus sacrificing the capability to analyze the consequences of structural change induced by energy policy. The model is not workable for policy questions related directly to refinery location decisions and other questions of the economics of the refinery sector; the model's rigidity and lack of environmental detail raises the possibility that the model might misrepresent the regional impact of policy changes on the refining industry.

Analysis of the model's behavior on the computer indicates that the model is "stable" only for a given range of product output/regional mix variations. One must know before hand whether the solution will be within bounds; the practical importance of this restriction is that these ranges are sufficient for short-term analyses, but not for longer terms where refining equipment can be changed by the industry.

Study 8 deals with the questions of validity, verification, workability, and clarity in MEFS's representation of the electric power sector. The representation of the electric power sector operation and behavior concerning the choice of fuels for generation, the capital structure, base load requirements, and other factors in response to changing rate structures and fuel costs are examined.

The basic questions addressed in this study were (1) whether the model representation of the electric power sector corresponds to current theory relevant to that of regulated utilities (validity), (2) whether the physical relationships concerning conversion efficiencies from various fuel mixes and changing base load requirements over time are maintained for a range of relative fuel mixes and growth rates for the sector output (verification), and (3) whether the sector representation is capable of answering relevant questions within the context of the overall model (workability).

In light of the evaluation criteria, the study makes three observations. First, the electric utility representation in the MEFS integrating model is the simplest possible. The only additional simplifications that could be made are to reduce the number of plant types or to reduce the number of load categories. Either would sacrifice appropriateness as an electric utility representation.

Second, the model will forecast behavior very poorly if unconstrained. However, the model has been used for the purpose of accounting, not for determination of behavior. Future expansion plans and plant utilization possibilities are entered as a very tightly constrained set of data. The model then is used simply to translate one set of prices (fuel and capital) into another (electricity).

Third, as an analytical tool in and of its own right, the model offers nothing that could not be obtained in more detail elsewhere. Thus the conclusion that the electricity sector of the MEFS model serves its intended, although somewhat limited, purpose adequately.

Study 9 deals with the questions of validity, verification, workability, and clarity in MEFS's representation of the various energy transportation modes. The representation of rail, barge, and pipeline transportation networks for moving basic energy commodities is examined. The transportation system representation deals with movements of coal, crude oil, natural gas, and refinery products at the wholesale level.

The basic questions addressed by this study include (1) whether the behavior of regulated transportation modes corresponds to current theory (validity), (2) whether physical relationships of capacity, time constraints on new capacity, and required product flows are maintained over a range of national and regional variations in the mix of energy production and consumption (verification), and (3) whether the sector representations are capable of answering relevant questions within the context of the overall model (workability).

Several major conclusions summarize the study results. The regional supply-demand structure of MEFS, using centroids within regions to represent the weighted centers of activity of different types, is an acceptable method to use in modeling energy distribution in the United States. However, several of the regions, such as the refinery region for the south-southwestern U.S. and most demand regions, probably are too large. In the case of coal, it appears that an attempt was made to correct for inaccuracies in the model basically caused by large regions by constraining the model solution.

MEFS's transportation models use constant transport costs for projection purposes. This assumption probably overstates the cost of transporting by pipeline and limits the model's usefulness for identifying changes in transportation modes over time. The forecasting method should have considered explicitly the trends in real transport cost by mode, thus raising a question of validity.

The transportation models used in MEFS have not been verified either *indirectly,* through verifying the overall MEFS model energy flows, or directly, through determining how well the models predict transportation costs for different material/model combinations.

Although the concepts used in the transportation models in MEFS are easily understood, many of the data inputs are inadequately documented, which considerably limits the full understanding of these models by users outside DOE. This lack of complete documentation, together with the lack of verification, either of the overall MEFS model or the specific transportation models, leads to the conclusion that MEFS cannot be relied upon by outside users at this time to evaluate important energy policies heavily influenced by transportation questions. High priority should be given to structuring the model in such a way that it can be verified using recent data. Also the transportation models should be reformulated to: (1) explicitly consider trends in transport costs by mode, (2) consider alternative methods of modeling coal transport among regions, possibly using additional demand nodes, and (3) use more definitive models for estimating costs of new or upgraded pipelines and rail lines.

Demand and Macro/Micro Interface Studies (Studies 10 and 11)

Three important modeling areas are crucial to the representation of the energy sector (and to energy policy effects) within the context of national and regional economies. First, the aggregate growth in economic activity has an important effect on the growth in energy consumption. Second, energy policy and energy sector behavior may have important macroeconomic effects or, at least, important effects on selected regional economies and economic sectors. Third, the direct consumption response to energy price changes and the substitution among energy forms and nonenergy products and factors of production resulting from changing relative prices of energy are of great importance. Two demand related studies were designed to assess the importance of these questions.

Study 10 deals primarily with the questions of validity, verifiability, workability, and clarity of demand models estimated for MEFS. MEFS specifies detailed demand functions by region and type of fuel. Demand relationships are specified according to economic theory; parameters are estimated with statistical procedures using historical data; and the model is then used to make simulations of energy consumptions over time under alternative price paths and various growth paths for income, population, and value added in manufacturing provided from the DRI model.

The basic questions addressed by this study include (1) whether the demand model specifications adequately correspond to current theory (validity), (2) whether empirical parameter estimates are reliable (verifiability), and (3) whether the dynamics of demand growth corresponds in significant ways to empirical data outside the data used in parameter estimates (verifiability).

MEFS's demand model is composed of three submodels, all of which are econometric in nature. The first (the residential, commercial, and industrial submodel) is the most complex of the three, as these sectors accounted for about 73% of national energy use in 1975. From a validity criteria point of view, the *resulting* model specification seems plausible, but one cannot be certain since the *original* model specification is not documented or available, nor can one trace the rationale of the modelers from start to finish in order to ascertain the reasonableness of procedures. Thus, the validity of the model is questionable. The ordinary measures of statistical validity of the model parameter estimates essential to sound principles of statistical estimation—coefficients and standard errors—are not available with the results, nor were they apparently used by the modelers; hence, there was no way to assess the stability or reliability of individual parameter estimates for the variables chosen for the models.

Projections for this model were accommplished by modifying the estimating equations. Some variables such as heating degree days were dropped, and the distributed lag format was also eliminated before projections were made. Thus, the projection equation estimates had to be bench marked to a certain year to guarantee comparability with estimating equation results.

The second part of the demand model, the transportation submodel, was structured according to end use of the fuel. Hence there were, for example, equations for auto highway gasoline use, non-auto highway gasoline and diesel fuel use, rail diesel fuel use, and commercial jet fuel use. In all, 22 equations were employed to estimate demand for fuel for the transportation sector; and single equation estimations, using ordinary least squares with time series data were the main statistical tools used. The use of endogenous variables as explanatory variables in the estimating equation leads to regression bias in the parameter estimates, and the off-diagonal elements of the error convariance matrix should be examined closely to determine whether bias exists in equations so constructed. The t-tests, R-square, and Durgin-Watson coefficients were presented for these models and they appear to be satisfactory. Price and income elasticity estimates also appear to be reasonable. However, information on the regional demand for fuel by the transportation sector was not made available to the investigators. Thus, no evaluation of the workability of the model at the regional level was possible.

The third submodel, the demand for minor fuels, included the following relations: (1) the demand for natural gas, liquid gas, and coal in the raw materials sector; (2) the commercial sector demand for asphalt and liquid gas; (3) the industrial demand for petroleum gases and metallurgical coal; and (4) the residential and commercial demand for coal. Only relation (2) above included price as an explanatory variable; all other relations used industrial value added, time and lagged endogenous variables as independent variables. The ommission or price as an explanatory variable in these equations denies the basic relation between price and quantity in demand theory. The use of time and lagged endogenous variables seriously reduced the model's ability to correctly estimate sudden changes in energy use that are quite different from past trends, yet these types of changes could have been present in the middle seventies. Statistical evaluation of the quality of parameter estimates was not possible because of inadequate information provided to the investigators.

Sensitivity analyses were performed on the computer in order to determine the practical importance of variations in price elasticity estimates in the demand model. Several attempts were made to obtain reliable results from the model system so as to determine if changes in the parameters have major impact on the overall solutions from MEFS's integrating model. Instructions from EIA personnel were followed after questionable results were obtained. The intended sensitivity tests remain inconclusive.

Results obtained at this writing seem to imply the following. Target year demand information (quantities demanded) for various end use fuels are virtually fixed for the purposes of the integrating model solution, being determined as they are by DRI macro variable forecasts *and* prior year's estimates of "independent" variables through the lag structure of the demand model. That is, so far as the demand side of MEFS is concerned, 1985 (or 1990) overall integrating model solutions are determined by the *price path* of previous years for final fuel demand by fuel type by sector; market clearing prices for the integrating model in 1985 are not significantly affected by a 1985 period price quantity relationship. As a result, the overall MEFS results are heavily determined *a priori* by specification of the price paths.

Study 11 deals with the questions of validity, verifiability, workability, and clarify of MEFS representation of the interactions between the energy sector and the macroeconomy. The representation of the interaction is in terms of (1) market clearing energy prices from MEFS to DRI which satisfy equilibrium conditions in MEFS's forecast years, and (2) macro growth forecasts of personal income, employment, value added, and other macro variables from DRI to MEFS.

The basic questions addressed in this study are (1) whether the representation of interaction between the macro and energy models of MEFS correspond to current theory of macro and energy sector economics (validity), (2) whether the interaction produces dynamic stability in terms of market prices, effects on the macroeconomy, and consistent subsequent behavior of the federal government concerning basic macroeconomic policy (verifiability), (3) whether the conclusions of small macro effects of energy policy and large energy effects of macro policy are verifiable empirically, and (4) whether sector and regional economic impacts of energy sector changes as measured by MEFS are reliable on theoretical and empirical grounds. The study examines the extent to which MEFS is able to answer important questions of regional and sectoral impacts of energy policy (workability).

Major conclusions can be summarized as follows. First, regarding the DRI/MEFS interface, the structure of the linkage is seriously inadequate. The formal links from MEFS to DRI are the wholesale price index for fuels and two detailed prices and quantities, ignoring much regional and sectoral detail. This simple linkage raises serious questions of validity and greatly affects the model's workability, since one is apt to be mislead about the economic effects of energy market changes. Major adjustments to both models are done judgmentally by the modelers. There may be substantial regional, sectoral, or industrial effects of changing energy prices; but these effects are not readily discernible by the models. There may be considerable short-run dislocations at the national level that are not captured by the long-run equilibrium solutions, again limiting the models' workability from a user's point of view.

Second, with respect to energy-regional economy interface, regions develop in lock-step with the national economy. The system is too simplified to provide much meaningful regional analysis to policymakers concerned with regional costs and benefits of alternative energy policies. Workability is weak for regional analyses.

Third, with respect to energy-international economy interface, the attempts to model functional relationships for oil exporting and oil importing countries suffer serious shortcomings. These shortcomings cause much of the analysis to be judgmental, with specific results being dependent on predetermined values of critical variables.

APPROPRIATE USES OF MEFS

MEFS was first designed to evaluate the impacts of national energy policy alternatives relative to President Nixon's Project Independence program and was known as PIES. Energy production incentives through decontrol of wellhead prices of domestic crude oil and natural gas, RD&D incentives for new energy resource development, import fees and quotas for imported crude oil, and energy conservation programs were among the national policy instruments considered in the initial modeling work. The primary focus was on measuring the likely contribution of these major policy alternatives to the reduction of U.S. crude oil imports.

Once the system was developed the kinds of analyses performed were expanded to more detailed evaluations, including the effects of oil and gas pricing, Alaskan pipeline alternatives, boiler fuel restrictions on natural gas use, economic impacts of shortages, and more recently the impacts of President Carter's NEP. Several analyses of various parts of the NEP were completed during the 18 months of congressional action which followed during 1977 and 1978.

The evaluations completed by TNEMP imply that the use of MEFS for energy policy analyses has several major weaknesses. Given the background of MEFS's development it is not surprising that it did not live up to even weak forms of our evaluation criteria. MEFS fails the clarity test because documentation is poor, dated, or non-existent. Even with reasonably good cooperation from DOE/EIA personnel, it was difficult for TNEMP evaluators to be certain that relatively simple parameter changes in the models were done correctly.

Validation tests illustrating the consistency of the model's behavior with underlying logic, theory and expectations of the modelers are virtually non-existent. TNEMP evaluators completed a number of sensitivity tests, but we could not claim to have verified any of the models. One variation of the finding rate in the oil and gas model produced an unbounded solution in the integrating model not before encountered by DOE/EIA, indicating a lack of verification testing by the modelers. Individuals within DOE/EIA undoubtedly have a better understanding of the model's behavior than TNEMP; but such understanding is not published, and the turnover of personnel makes such information difficult to transfer even within DOE/EIA.

Although MEFS is operationally very complex, the underlying conceptual framework is quite simple. In fact, the model is simplistic on several counts, limiting its usefulness for certain important policy questions on the basis of validity tests. The oil and gas supply models (1) use a very simple formulation of the industry investment behavior with no empirical work to verify the formulation, (2) ignore the basic joint product relationship between oil and gas, and (3) are inconsistent as between the methodology for specifying the finding rates for oil and gas and the economic specifications of the models.[8] Reasonable variations in finding rate specifications for the model produced great variation in the price elasticities of production in the price range important to recent policy decisions. In particular, when the finding rate was reduced by 50% (near that suggested as more appropriate by TNEMP studies) the price elasticity for gas increased from 0.200 to 2.320. Since a major policy issue under congressional action on the National Energy Act was the price responsiveness of gas production at prices of $1.75 and above, use of this model in the discussion may have contributed greatly to confusion. The choice of a high finding rate, necessarily a very judgmental input to the model, produces a very low (0.200) price elasticity of supply in 1985, whereas a 50% lower finding rate shows a relatively large (2.320) price responsiveness.

Evidence from TNEMP studies indicates that cash flow and source of investment funds are significant factors in oil and gas investment decisions, in addition to the marginal price. Since the oil/gas supply models rely only on a simple investment model formulation of industry behavior as a function of marginal prices and costs, they are of limited usefulness for oil pricing policy analyses now a major issue.

In short, MEFS's oil and gas supply models are of limited usefulness in answering the major policy questions of how much oil and gas production increases can be expected from domestic sources under various pricing policies. Answers from the models will vary greatly under different assumptions made as judgments of the modelers. The models are too simplistic and in need of a more sophisticated conceptual framework with empirical content. The models may be as useful as other available oil and gas supply models for general outlook work understood to be heavily dependent on EIA personnel judgments but should not be relied upon heavily for detailed policy analyses of oil and gas pricing, taxing, and allocation.

The demand model has major weaknesses limiting its use for policy analyses which depend on estimates of the rate of growth in overall energy consumption or on measurement of the ability of the economy to make fuel substitutions in response to relative price changes. The model seems conceptually sound (as a generalization) when compared to economic theory; i.e., the theoretically most important variables to explain energy consumption are included in the model. However, the methodology used to interface the dynamic demand model with the integrating LP model to simulate market clearing conditions seems confused. The formulation of functional relationships between price and the quantity of various fuels taken off the market at a point in time do not correspond to the concept of a static demand curve understandable to economists. Further, it is doubtful that available consumption and price data will support the recent versions of the model. The usual test statistics (used by econometricians and statisticians) are not available and apparently were not used by the modelers to determine the reliability of parameter estimates derived for the model. Thus, one has little basis for confidence in the model behavior.

DOE/EIA now has a data gathering exercise in progress which should enhance the quality of the demand model, but these data will not be available for some time. In the interim it will be difficult to establish the demand models reliability without reestimating the equations, documenting the rationale for model modification and development, step by step, and the provision of test statistics so that outsiders can judge the reliability of the parameter estimates.

Sensitivity analyses of the demand parameters (own and cross price elasticities) by TNEMP evaluators led to the conclusion that target year (1985 and 1990) solutions of MEFS are not sensitive to price changes in those years; demand levels are primarily determined by *a priori* price paths specified by the modelers. It is unclear how one can specify *a priori* price paths to drive the demand model, when market clearing prices are supposed to be a major output of the modeling system.

Given adequately specified supply and demand model specifications, MEFS's integrating model will provide adequate information on the national market clearing prices and quantities of energy resulting from a variety of energy policy specifications. Calculations of import levels as the differences between domestic demand and supply will necessarily be more subject to variation (and therefore less reliable) than individual, major energy production and consumption estimates since imports are subject to error from two sources simultaneously.

The interface between MEFS's integrating model and the DRI macroeconomic forecasting model is accomplished through effects on the wholesale price index for fuels and specific price/quantity solutions for two fuels. Any other implications from the integrating model solution, such as crude oil import levels, are changed as *ad hoc* adjustments to DRI. Impacts of regional and sectoral investment shifts as a result of energy policy changes are not measured by the model interface. Effects of energy policy changes on the U.S. inflation and balance of trade are at best crude measures and depend heavily on the judgments of DRI modelers concerning international trade adjustments and U.S. fiscal/monetary policy response to the energy policy change. The DRI model is likely to be much more accurate at measuring inflationary impacts of U.S. monetary/fiscal policy changes than energy policy changes. In short, MEFS/DRI systems operated together will provide useful, partial analysis information on macroeconomic impacts of energy policy changes but should not be expected to provide adequate measures of long-term impacts on U.S. inflation, balance of trade, balance of payments, and the value of the dollar.

The MEFS/DRI/BLS Regional I/O systems operated together will not be very helpful in measuring the U.S. regional impacts of energy policy changes except in the very short term, one to three years. The models do not capture the changing regional investment flows and costs of living impacts which result from relative energy price changes, energy availability, and costs of doing business resulting from national energy policy. MEFS's integrating model will not be very helpful in addressing the impacts of policy on refinery location decisions and the regional economic shifts brought on by the substitution of coal, due to limitations of the refinery sector and transportation sector specifications.

RECOMMENDATIONS FOR MODEL IMPROVEMENTS

Oil and Gas Supply Models

MEFS oil and gas supply models as currently structured and documented should not be used for policy analyses and EIA has wisely recognized this as they are now planning to replace the models.

1. Considerable empirical work should be done to model the investment behavior of the oil and gas industry paying attention to the factors identified in the TNEMP studies, namely, the incorporation of risk and the inclusion of the source of funds and cash flow variables in an investment model.
2. Careful attention should be given to the methodology for specifying finding rates. Studies should be commissioned to separate economic and geologic information embedded in the data one must use to calculate finding rates. New, more detailed studies to characterize the frequency, concentration, depth, and quality of the oil and gas resource base are needed.
3. Variations in supply response as influenced by alternative finding rates, discount rates, rig life, and other parameters, should be incorporated with analyses. Such procedures will provide the user with much better information about the reliability of projections.
4. Empirical work should be done for the oil, gas, and coal supply modeling efforts to determine the influence of various regulatory procedures on investment behavior. Such information will be of growing importance as we apparently can expect such regulation to be pervasive in these industries for the foreseeable future.

Demand Model

The demand model should be reestimated clearly documenting the modelers' decisions in model reformulation throughout the estimation exercise leading to the specification of the final model.

1. All statistical tests should be made available, so that outsiders can evaluate the soundness of model parameters for projection purposes.
2. The dynamics of the demand model and its interface with the supply models in the integrating LP model at a pont in time is confused and needs clarification and further work.
3. The practice of interrelating econometric demand models with engineering process supply models raises many methodological issues. Consideration should be given to the use of an engineering process modeling approach to demand modeling which would be consistent with the methodology employed elsewhere in MEFS.

Energy/Macro Model Interface

The current linkages between MEFS's energy market model and the macroeconomic models is very weak.

1. Work should be done to identify the macroeconomic impacts of changing investment/disinvestment and regional/sectorial investment flows resulting from energy policy decisions. Major shifts in energy production among coal producing regions and among oil/gas and coal

producing regions will obviously accompany major energy/environmental policy decisions. Location decisions for major energy using and processing industries will also be affected. Such shifts in economic activity are *not free* compared to alternative policies, and such cost should be captured by the models.

2. Better modeling to identify regional environmental impacts of energy/environmental policies should be undertaken. These regional, economic, and environmental impacts are of major economic, environmental, and political importance, and the arguments over these issues are perhaps *the* major impediment to achieving a consensus on national energy policy.

RECOMMENDATIONS FOR MODEL TRANSFER

1. Major efforts must be made to achieve a "living" model transfer of MEFS. (See also the "Highlights" section and the Advisory Board recommendations.) An archival system as now planned by EIA, though helpful (if timely), is not adequate to achieve the "spirit" as well as the mechanics of model transfer. Hopefully, increased attention to problems of poor model documentation will lead to more transparent models, but such documentation is not now available making mechanical transfer almost useless.
2. A major finding of the TNEMP is that the large complex modeling system heavily laden with judgments of the modelers cannot be meaningfully transferred in the absence of major interpersonal contacts between the modelers in EIA and the model recipients. EIA should make arrangements for continuing assessment of the TNEMP type. EIA plans indicate major evaluation exercises on a five-year interval, but such an approach avoids important interim interaction of the TNEMP type.
3. Much more attention must be given to verification work by the modelers not only to improve their own understanding of how the model behaves but for use by outsiders to make judgments about the reliability of model results. Such information is vital to model assessment by third parties.

FOOTNOTES
CHAPTER 2

[8] Conceptual and empirical work needs to be done to separate the economic and the geologic information that influence the observed economic reserve additions from incremental drilling in historical data that forms the basic data for finding rate estimates.

CHAPTER 3

FURTHER MODELING ASSESSMENT AND DEVELOPMENT FOR TEXAS

by Milton L. Holloway

INTRODUCTION

It is well recognized that national energy policy decisions are of great importance to the future growth and environment of Texas. The Texas National Energy Modeling Project (TNEMP) has documented the need for better analytical information on expected Texas impacts of national energy policy decisions, and this chapter makes recommendations for a modeling development program to provide such information to the Texas citizenry and political leaders responsible for setting state policy on energy and for making recommendations on energy policy matters to their counterparts in Washington, D.C.

TNEMP also provided valuable insight into the best institutional arrangements and procedures for ensuring that analytical results from modeling and data base systems are as reliable and useful as possible. The most important attribute of such an information base is workability, that is, the applicability of the modeling system to practical problems. The success of such applications will be greatly dependent on institutional factors. The implication here is that model *development* and *analytical studies* must be done with a keen awareness of perceived problems of the user.

In order for study results from such modeling efforts to be accepted, five important conditions must be present in the following order of importance: (1) the study must be well-timed; (2) the results must be effectively communicated by a trusted aide of the policymaker in terms the policymaker can understand; (3) there must be compatible personalities at the policy interface; (4) the modelers must be relatively independent of the policymakers; and (5) the modelers must have a reputation for astute analysis and good sense based on past performance.[9]

The implication here is that personalities, interdependence, independence, and professional reputations are involved in making the study results transferable and useful. Problem definition and the translation of study results must be from "close in" to the policymaking process; the analyses must be one step removed from the policymaking process but timely and in close working relationship to the individuals involved in the policy process. In addition, institutional arrangements must be formalized for third party definition of standards for documentation including procedures for access. A third party institution should also provide for performance evaluation of the modeling system relative to identified uses and purposes.

TNEMP has enhanced and further cultivated a working relationship between TEAC and the major Texas universities. Most of the best known and respected Texas professionals in various areas of energy expertise have been involved in TNEMP. Formalizing this working relationship will provide for an ongoing investigative capability to develop and use models in the TEAC policymaking process. The following sections define working relationships, provide for documentation and evaluation, and establish a rough work plan for the first two years' operation.

TEXAS ENERGY POLICY PROJECT OBJECTIVES

The ongoing work will have an expanded purpose and therefore may be more appropriately named the Texas Energy Policy Project (TEPP). The primary objective of TEPP is the integration and further development of Texas energy investigative capacity for use in TEAC's decision-making process. A considerable capacity already exists in Texas but its components reside in separate institutions and are in need of coordination for TEAC policymaking purposes. In addition, certain weaknesses exist in the present Texas investigative capacity and these gaps must be filled to enhance the quality of decision-making. Furthermore, new technologies of computer and mathematical modeling not heretofore available make possible the identification of impacts and the improvement in understanding of complex relationships which must be exploited for purposes of TEAC decision-making.

Since Texas energy policy decisions faced by TEAC are so heavily influenced by national policy, it is important to interact with the Federal government in its policymaking decisions, both at the investigative and political levels. By maintaining a current working version of DOE's primary mathematical model in Texas, communications are enhanced at the investigative level. A dialogue among investigators is encouraged and such communications enhance the performance and understanding of both groups of investigators.

The following strategies should be pursued through the operation of TEPP to achieve the primary objective:

1. Identify more precisely the investigative units and capacities in Texas easily available to TEAC.
2. Identify more precisely existing formal mathematical models in Texas of the state and national economy, population growth, energy resources and production, and environmental quality.
3. Identify and continually reassess the major gaps in investigative capacities and formal mathematical models.
4. Develop a mathematical modeling system which makes maximum use of existing models while formalizing and further developing a modeling system to characterize the influence of energy policy and energy

development on the economy and environment of Texas.

5. Formalize the integration of existing investigative capacity and newly developed capacity with the TEAC decision-making process.
6. Maintain continual interaction with DOE's investigative units through maintenance of the federal government's primary mathematical modeling tool.
7. Produce an annual assessment of DOE's national energy outlook (the Administrator's Annual Report and the Policy and Evaluation's National Energy Plans).

MAJOR CONSIDERATIONS FOR STRUCTURING TEPP

In order to structure TEPP to accomplish the objectives outlined in the previous section, certain Texas institutional relationships and the character of the decision-making process of TEAC must be understood. This section summarizes such material so as to represent the general principles of decision-making and the existing structure of Texas institutions.

Understanding the Structure of TEAC Decision-Making

A decision-making unit is characterized by various activities and the accumulation and use of various kinds of information. Figure 3 is a diagram of the decision-making process in general.[10] Such a process is common to decision-making units that have the power to decide and act while being required to bear responsibility for results.

As shown in Figure 3 the decision-making process is very interdependent. There is a continual process of integrating information and obtaining feedback. The decision-making unit integrates normative and positive information, makes a prescription for action (including a decision to do nothing) and bears responsibility for the results. The process requires both administrative and investigative capacity. Problem definition falls in both administrative and investigative capacities; observation and analysis-synthesis fall mainly in the investigative capacity; and decision-making, action, and responsibility-bearing fall primarily in the administrative capacity. Various individuals or a single individual may be involved. Typically, such a decision unit invovles an organization of policymakers, advisors, and analysts who perform somewhat separable functions.

The decision-making process begins with a definition of the problem under consideration. This definition must come jointly from the administrative and investigative capacities but must be well understood and endorsed by the administrative group since they ultimately will be responsible for decisions and will bear responsibility for the results. In the case of TEAC, problem definition is the joint function of the Council and the staff. Problem definition is not usually a single action but rather is a

Figure 3

The Structure and Components of Decision Making

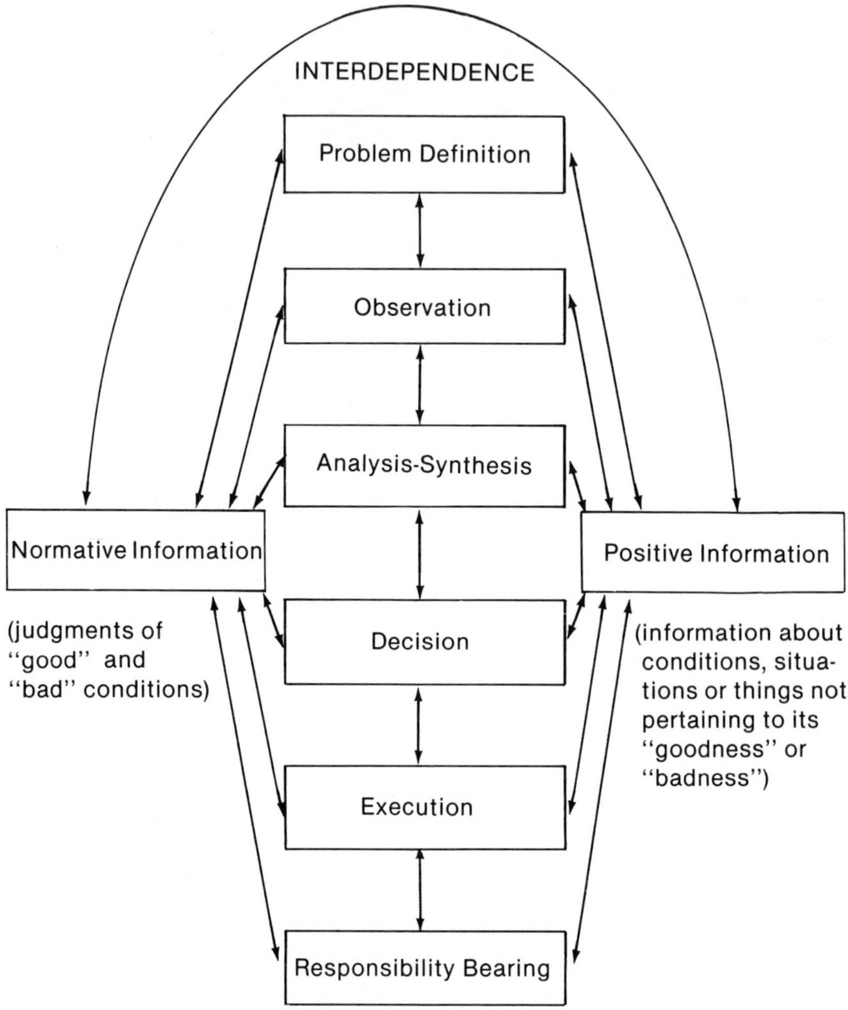

Adapted from Johnson, et al, *Agricultural Sector Planning: A General System Simulation Approach,* edited by George E. Rossmiller. Agricultural Sector Analysis and Simulation Projects, Department of Agricultural Economics, Michigan State University, East Lansing, Michigan, 1978, p. 34.

sharpened conception as feedback occurs throughout the process. Conceptualization of the problem requires reliance on several disciplines (e.g., engineering, economics, geology, etc.) and subject areas (technical, institutional, and humanistic aspects of the problem).

The second function in the decision-making process is observation. This function includes collecting and processing various kinds of data and information. Important information is likely to be needed from political, social, economic, and physical systems. The determination of what information should be collected is indicated by the problem definition. However, limitations on how much information is to be collected, since this function is costly, is determined by judgments of the investigative unit interacting with decision makers. Cost and timeliness will be important in answering these questions. In the case of TEAC, this is a staff function.

The third function in the decision-making process is analysis and synthesis. This is the investigative unit's function and may involve the development and use of formal mathematical models of the economy, the environment and other systems, and may make use of disciplinary models (economic, geologic, environmental); it is rare, however, that a real world problem ever involves only one discipline (economics) or one subject matter (energy supply industries). This function often involves the use of other studies and outside experts to provide breadth and depth to the investigation. In the case of TEAC this is the function of the staff, although it should be abundantly clear that Council members have access to and will make use of other investigative units at their disposal.

The fourth function in problem solving is decision. In this function all of the relevant information (data, theories, and values) is brought together to form a prescription for action. The conceptualization of this function is that it provides the best decision available, though such a decision may often be making the best of a bad situation. In the case of TEAC, this is a Council function.

The fifth function is action or execution of the decision. This is a Council function in the case of TEAC. Council actions may be to give advice or admonition to others to act since the Council is an advisory body to the Legislative and Executive branches of Texas state government. In other cases, the Council acts on energy development and conservation plans and projects.

The sixth function is responsibility-bearing. Responsibility is borne by decision makers, action takers, and affected people. In the case of TEAC those affected by energy policy have the power to provide feedback to Council members and therefore participate in Council deliberations. This feedback is provided by Advisory Committees of lay people from the Texas citizenry. This is a particularly important part of the decision-making process keeping problem definition and problem solutions relevant to real world problems.

Overview of Texas Investigative Capacity in Energy

Texas state government is a decentralized form of government having many functions performed under the direction of independently elected officials. Important independently elected Texas officials providing a portion of the State's investigative capacity in energy in addition to the Council staff include the Governor, Lieutenant Governor, Comptroller, Attorney General, Commissioner of the General Land Office (manager of state-owned lands), Commissioner of Agriculture, and the three commissioners of the Railroad Commission. There are also a number of important agencies directed by commissioners or boards appointed by the governor who provide energy investigative capacity. These include the Public Utility Commission, the Air Control Board, and the Water Development Board. The State's universities also provide a major source of energy investigative capacity.

The chief access to the university energy investigative capacity is through various energy institutes established to coordinate energy research on campus. The principle organizations include the Energy Council and the Center for Energy Studies, The University of Texas; the Center for Energy and Mineral Resources, Texas A&M University; The Energy Lab, University of Houston; and the Center for Energy Research, Texas Tech University. Other institutions having important investigative capability include the Bureau of Economic Geology and the Bureau of Business Research, the University of Texas; the Texas Transportation Institute, Texas A&M University; the Texas Research League; and the Legislature Budget Board. The House and Senate energy committees also provide investigative capacity.

The character of investigative capacities in each of the above institutions except for the university energy centers, are designed to meet the decision-making needs of the institution in which they reside. None of the institutions have overall responsibility for energy policy planning as is the case for TEAC. And, the university centers are not doing problem solving research in the sense of the last section. All of these institutions will have information useful to TEAC, but cannot substitute for the investigative requirements of it. Furthermore, it should be clear that individual Council members will make use of one or more of these investigative units to provide information for his participation in TEAC, in addition to, or irrespective of information from the TEAC staff.

Texas independent state agencies listed above each have unique statutory functions differing in both substance and planning horizons. It is important to understand these differences when examining mutual interest and joint working relationships in the provision of investigative capacity for the decision-making of TEAC.

The Governor is by law the "chief planning officer" in the state for purposes of federal planning monies coming to the state. These planning functions are carried out by the Governor's Budget and Planning Office (GBPO) or other offices selected by the Governor. GBPO focuses on budget planning for the short term (the next biennium) and for the long term in other special assignments. The Governor has long and short range energy policy planning functions.

The Lieutenant Governor, through his position as chairman of the Legislative Budget Board (LBB) and chairman of TEAC, has both long term and short term interests and responsibilities for state planning. The focus of the LBB, however, is mostly short term extending to the end of the next biennium. The focus of TEAC is generally long term (five to twenty-five years) since energy supply and demand responses and development programs are long term problems.

The Comptroller's statutory responsibility is the certification that state budgets approved by the legislature are equal to or less than his revenue forecast. Although the Comptroller has an interest in long term economic forecasting, his primary focus is on the two year planning horizon commensurate with his statutory responsibility. Energy production, gasoline consumption, and energy prices are of great importance since they are important sources of state revenues. The Comptroller has considerable investigative capacity in revenue forecasting.

The Attorney General does not have a specific planning function but makes use of energy analyses and forecasts useful in the preparation of case materials in his role as intervenor in legal cases for the state. The Attorney General has considerable investigative capacity in legal matters of energy policy. The Commissioner of the General Land Office has a specific need for the information concerning the outlook for oil and gas reserves, production, prices, etc. since he is responsible for the leasing of state-owned lands; his planning horizon is both long and short.

The Commissioner of Agriculture has no specific planning function in agriculture but has investigative capacity in agricultural energy production potential and impacts of energy cost and shortages on Texas agriculture. However, he has no statutory responsibility specifically requiring economic or energy forecasting and analyses.

The Texas Public Utility Commission has major investigative capacity in electric utility rates and financial analyses of electric utilities, focusing on the short term, to carry out its regulatory functions. The PUC has an economic research unit capable of using and contributing to modeling and analysis efforts in electric utilities; however, time and resources have not allowed a long term effort to date.

The Texas Railroad Commission has considerable investigative capacity in the area of oil, gas, coal, and uranium regulation, focusing heavily on legal research. Since the oil and gas regulatory function is based on a statutory responsibility of "conservation" of the petroleum resource, the Commission has not been historically interested in price information and does not have a staff function dealing with forecasting.

The Texas Department of Water Resources Board has considerable investigative capacity in areas of water resource planning, including state and sub-state level planning of water development for energy facilities. The Board has a long term planning horizon ranging up to 50 years into the future.

The Texas Air Control Board (ACB) has investigative capacity in air quality impact assessments in accordance with its statutory responsibility to enforce the provisions of the Federal Clean Air Act. The ACB's focus is mostly short term and is concentrated in environmental engineering and related disciplines.

The four energy institutes, in addition to the Bureau of Economic Geology, the Bureau of Business Research, and the Texas Transportation Institute, have disciplinary and subject matter investigative capacities in various areas of emphasis. The Bureau of Business Research is the most active state institution involved in the provision of general economic outlook information, both for research purposes and for use by the business community in their planning activities. The Bureau of Economic Geology is the state's geological survey function, providing information and analyses of geological resources. The Transportation Institute focuses on transportation problems in Texas, emphasizing highway and rail transportation research. The Center for Energy Studies has major strengths in electric utility economics and engineering, geothermal resource development and local impact assessment. The Energy Institute has major strengths in large scaled solar applications and economic modeling. The Center for Energy and Mineral Resources has major strengths in lignite and biomass technologies. The Center for Energy Research has major strengths in biomass and wind technologies.

The Role of Formal Mathematical Models

Mathematical models serve several functions in the policy process. They are useful not only for resolving issues but for raising issues. Models are also used to justify or reinforce decisions or predictions already made.

In the problem solving process described in a prior section it is clear that formal models become a tool of the investigative capacity of the decision making unit. Perhaps the most common use of models in this context is to forecast or predict the effects of alternative courses of action. Such forecasts are "conditional" forecasts, initiated to provide information about the difference in the system performance under conditions of the proposed policy decision. Models only provide information about selected indicators of system performance and never provide complete information. Informal models, opinions, data, historical experience, and advice must round out the information requirements for decision-making.

The policymaking process is not only concerned with resolving issues, however, but is also concerned with raising issues. Therefore, formal models are useful for converting conceived problems into political issues by drawing public attention to the topic, assembling support for proposal actions, resolving conflicts, legitimating decisions, and securing compliance.[11]

Major disappointments in the usefulness of expensive models built for policy purposes can be explained in many cases by referring to the decision making structure described in the previous section. It is clear that problem definition must be the joint effort of both the administrative and investigative capacities of the decision-making unit. If the administration—the decision maker—does not agree with the problem definition, the subsequent investigation which involves the construction and/or use of models, will be misdirected, from the decision maker's perspective, providing irrelevant or misguided information. In the case of TEAC, the Council, or at least key members of the Council, must be involved in problem definition if formal models are to be useful. Furthermore, the values of the decision maker must be understood and respected in the modeling process so the results can be useful; for example, a model constructed to provide guidance to a government agency, presuming the government ownership and operation of the oil and gas production industry in the country, will not be appreciated by a policymaker who believes in free enterprise.

Formal models and model results must also be timely if they are to be useful in the policy process. Issues arise from many quarters and often must be addressed quickly. The best available information must be brought together rapidly. The ability to use formal models in such circumstances is limited. Existing analyses may be useful but are seldom definitive. A close working relationship between the administrative and investigative capacities, however, helps identify upcoming issues and provides enough lead time for formal modeling.

The work plan section identifies major existing, formal models in Texas useful to TEAC decision-making. The work plan also establishes a process for formalizing a modeling structure to make use of models, both inside and outside TEAC, and to construct new models in order to enhance the investigative capacity of TEAC.

Working Relationships for TEAC and Texas Universities

Major Texas universities have considerable expertise needed in energy policy analyses for TEAC. The disciplinary focus of university departments and colleges provide essential depth of knowledge in areas such as geology, engineering, economics, law, political science, and sociology. Such expertise is developed over long periods of time for the primary purpose of education and research to preserve and further develop the discipline. Therefore, university departments are strongly structured for disciplinary research, not for subject matter research, and certainly not problem solving research. Other organizational arrangements must be created to build subject matter research and problem solving research capabilities.

There are in existence many on-campus centers created by the universities to focus on subject matter research. For example, there are water resource institutes, transportation institutes, population research centers, and energy centers or laboratories. These centers, institutes or laboratories coordinate, organize, and focus multidisciplinary research efforts into subject matter areas. Each of the four major Texas state universities has one or more such organizations in the energy subject matter area. These are (1) the Energy Council and the Center for Energy Studies, The University of Texas at Austin; (2) the Center for Energy and Mineral Resources, Texas A&M University; (3) The Energy Laboratory, University of Houston; and (4) the Center for Energy Research, Texas Tech University.

Further organization is required, however, to transfer this disciplinary and subject matter research capability into energy problem solving research capability. This task can be accomplished to take advantage of the best capability on the several campuses only by direction and organization from an outside group that forms a decision-making unit. TEAC has such a capacity.

Since TEAC is interested in and responsible for energy policy formulation for Texas, and has the power to make such decisions as well as the responsibility for accepting the decision consequences, it constitutes the primary energy decision making body of Texas state government. Furthermore, since the chief executive and legislative officers of Texas sit on the Council this organization has considerable influence in guiding university research work to focus on Texas problems; conversely, TEAC defined problems are the natural focus for university researchers interested in providing research results which bear on current problems. Therefore, a natural, compatible working relationship can be created if properly organized. Such an organization can transform the disciplinary and subject matter research focus already well organized on the four university campuses into problem solving research needed by TEAC.

Two important and necessary conditions are present in TEAC. First, adequate funding exists to pay the marginal university costs for model development, individual research projects, and travel for meeting participation by key individuals. Second, TEAC has a competent junior level professional staff for assisting in the Council decision-making process by identifying current issues, tracking energy policymaking elsewhere, and transforming research results into timely papers and policy recommendations for consideration by the Council.

TEPP STRUCTURING AND FUNDING

The working relationship among interested parties must be formalized and funding must be provided. This section establishes the organizational structure, identifying the roles of each. Funding to support the work is also identified.

Institutionalization of TEPP

TEPP should maintain three ongoing functions. First, a continuing assessment of the Energy Information Administration's (EIA) modeling system is needed to make judgments about strengths, weaknesses, and needed improvements in their modeling system. This activity should take the form of an annual review of EIA's Administrator's Annual Report to Congress as well as National Energy Plans as appropriate. Second, TEPP should become one source of immediate access to EIA's models as a service to the rest of the nation which would do much to fulfill their legislative mandate in the area. EIA would be encouraged to pay for this effort. Third, Texas needs to better develop our own modeling capabilities.

The needed modeling development for Texas energy policy evaluations involves two major parts. First, the key relationships of the Texas economy to the national economy must be captured in a far better way than currently exists in in-state modeling efforts or in national private consulting modeling available under contract. This general economic forecasting capability can be of general use to several Texas institutions and should consist of U.S. and Texas macroeconomic forecasting. Second, the energy sector and the environment must be modeled and interrelated to the overall economic model; several energy models currently exist in the state and should be examined and used or further developed where appropriate.

The two modeling counterparts should be developed in the newly created TEPP established for the purpose. TEPP should serve as an umbrella for current experts available in Texas state agencies, universities, and private consulting firms to develop the models. Once the development, testing and documentation work is complete, the models would be available for transfer and use by TEAC.

An interagency advisory board should be appointed to ensure that the modeling development incorporates the major features needed by their respective agencies as participants in TEAC. The national advisory board concept used in the TNEMP should be maintained for advice and guidance based on prior developments elsewhere. Ad hoc committees from

TEAC's Advisory Committee should be formed for each major policy issue to be investigated. Interaction with these groups is necessary to ensure realistic policy prescriptions since they represent the values of individuals, firms, and institutions affected by policy.

TEPP should be maintained for purposes of public access and transfer of the models and supporting data bases.

Two separate institutions should be established by the Texas Legislature for purposes of writing standards for documentation and transfer and for evaluation of model performance. Standards for documentation should be developed by the Systems Division of the state Auditor's Office; the Systems Division already works with the Governor's Office to review and approve computer systems rental and acquisition, including software. The model performance evaluation should be done biennially by the Legislative Budget Board consistent with its budgeting and program evaluation responsibilities.

TEPP As a Consortium

A formal statement of cooperation called a "memorandum of understanding" should be signed by the four university Presidents and the TEAC chairman. Such an instrument should be accompanied by interagency contracts from TEAC to each university to cover travel costs to Austin for occasional meetings and up to two days per month consulting time for key individuals on each campus. Contracts for model development and individual policy analysis studies should be awarded on a competitive basis as needed to support the policy work of TEAC.

TEAC Executive Committee

An Executive Committee of TEAC formed to expedite decision-making for the Council should serve as the primary interface of TEPP with the decision-making portion of TEAC. Meetings should be structured as needed to identify problems, determine value definitions, and review investigative results.

Early meetings of the Executive Committee should be required in the development phase of TEPP. Follow-up meetings should take place as new problems arise and research results and policy recommendations are obtained.

National Advisory Board

In order to provide access to guidance from nationally recognized experts in energy policy and modeling for policy analysis, a small group of individuals should be retained to be known as the National Advisory Board (NAB). This group can provide advice on model development and policy analyses, reference to experience and work of others, and an outlet for

Texas modeling and policy analyses to other analysts and policymakers in the nation.

NAB members should be retained under contract to TEAC for travel and consulting fees for several one to two day meetings during the year. Individual members could change due to the varying development phases of the work and changing personal interests.

TEAC Advisory Committees

TEAC's statute calls for the creation of a citizens advisory committee to be made up of a cross section of social-economic, geographical and special interest groups. This committee is to provide advise to the Council on policy issues. Historically this committee has been formed as a standing committee subdivided into functional interest or issue areas of energy. Other ad hoc subdivisions are formed from time to time as issues arise, and additions to the advisory committee are sometimes made to include other interests as appropriate. The structure of TEPP should include interaction of the investigative unit with such an advisory committee formed to address specific major issues. This interface provides a source of pragmatism for policy recommendations, and helps assure that the investigative unit works on relevant problems.

TEPP Analysis Team

A team of individuals should be formed made up of representatives from the four universities and the TEAC staff. Two groups, not necessarily mutually exclusive, should be identified and retained under contract to TEAC.

First, a peer review group (PRG) should be formed to evaluate and critique the Analysis Team procedures and research results and make proposed policy prescriptions to the TEAC Executive Director for consideration in decisions by the Executive Committee. Such prescriptions will necessarily be from a disciplinary perspective. The research findings of the second group should be reviewed and critiqued formally so as to guarantee improved research work. PRG should be coordinated by the TEAC Executive Director.

Second, a research group (RG) should be composed of a core modeling team (CMT) and policy research contractors (PRC). These groups could be retained under contract or as employees of TEAC. CMT should be responsible for formal model development and operation to characterize, as appropriate and to the extent possible, the expected impacts, costs and benefits of alternative policy actions. Modeling development by CMT should be a developing process, improving over time. The beginning work should make use of existing modeling capability available to TEAC, molding the pieces together into a consistent framework of flexible, general systems simulation and developing missing features important to TEAC decision-making. CMT should further utilize results from PRC to round out policy evaluations and to enhance development of better modeling capability. These two groups should be directed by the TEAC Executive Director.

Interaction of TEPP in TEAC's Operation

Having formalized TEPP as a consortium with individual members of AT and NAB identified work should begin by CMT and input should start immediately from PRG. With input from PRG, problems can be identified by the Executive Director for review by the Executive Committee. This review should formalize the definition of problems to guide the work of TEPP. At the same time the Executive Director should identify a set of basic values also to be reviewed by the Executive Committee and representatives from the Advisory Committee.

PRG should be available to respond to current problems and to make recommendations immediately, relying on existing research results and modeling capabilities. As the model development work is formalized, the results should be integrated into the process.

As problems are defined into various issue areas ad hoc groups will be formed from TEAC's Advisory Committee in order to ensure interaction with those impacted by policy decisions. The Advisory Committee forms a pool of individuals from various areas of the Texas citizenry from which to form such ad hoc groups.

The formalization of TEPP and all contractual arrangements should be completed during 1979. One or two meetings should be held with the Executive Committee to form a statement of basic values and problem definitions. As policy decisions are required by TEAC, PRG should operate to provide input as needed. PRG should also provide review of the modeling development plan for CMT. Once the proceses is fully developed, the Executive Committee should interact with the Executive Director and staff to define and re-define problems for analysis and policy recommendations. CMT should develop and maintain a flexible modeling tool designed to simulate the Texas energy sector, the Texas economy, and the Texas physical environment, with appropriate interrelationship to the national counterparts. PRG should provide critique and professional advice to ensure credible research results. The Executive Director and staff should translate the product of this group into policy prescriptions for Council consideration, providing feedback of effects from prior decisions from the Advisory Committee.

Funding

TEAC will fund TEPP for the first two years' effort under existing appropriations. Approximately $250,000 per year will be available for planning, development and testing work, and operation of TEPP.

The Department of Energy should be encouraged to fund continuing transfer work for EIA's MEFS for purposes of providing transfer of the "living" model making it accessible to the other users. This effort could further enhance the developing art of model transfer and assessment leading to more reliable modeling results and increased credibility for both EIA and TEAC.

SCHEDULE OF WORK FOR TEPP: 1980-1981 BIENNIUM

The first year's work would be somewhat unique in that TEPP would be functioning in an operational mode while at the same time, developing a new modeling system for use in policymaking.

Determination of TEAC Values

During the early months of operation, the TEAC staff should work through a process of formally identifying basic TEAC values to guide the work of TEPP. Working papers should be drafted by the staff with review by the Executive Committee and by the Peer Review Group.

The Definition of Problems

During the early months of operation TEAC staff should identify, with PRG and Executive Committee review, the basic problems to be addressed by TEPP. This process will have to be updated periodically since problems and perceptions of problems change over time.

Review of the State-of-the-Art in Regional Modeling

A review of the state-of-the-art in regional economic modeling should be completed in preparation for definition of model construction. A review of modeling computer languages ideally suited to TEPP should also be completed.

Utilizing Existing Mathematical Models in Texas

The University of Texas has three existing models of likely usefulness to TEPP. A U.S. regional electricity model with a Texas sub-part simulates the behavior of the electric utility industry. A population simulation model portrays the changing age, sex, and race characteristics of the population over time. A linear programming model of refinery transportation indicates the change in regional location of refinery capacity. New methodologies for characterizing coal resources is underway.

The University of Houston has a system of linear programming models of the U.S. refinery, chemical, primary metals, and electric utility sectors. U.S. and Texas oil and gas process models have also been developed but have not been recently maintained.

Texas A&M University has linear programming models of the agricultural sector of the U.S. and Texas economies.

TEAC has access to DRI's macro economic and energy annual and quarterly forecasting models through an annual contract. DRI could construct and maintain, with Texas input, a Texas data base and develop a Texas economic forecasting model appropriately linked to the DRI macro economic model.

TEAC has an annual Texas input-output modeling system which operates as an annual simulation model incorporating the effects of declining oil and gas supplies, rising energy prices, and energy substitutions by major energy using sectors. TEAC also has a Texas and U.S. natural gas market model and a commercial, industrial, and residential electricity demand model.

Further, the Analysis Team has prior experience with the DOE Midterm Energy Forecasting System maintained by TEAC at Texas A&M University and also has some experience with the DRI modeling system.

The Texas Department of Water Resources maintains a 1972 input-output model of the Texas economy. A 1977 version of the model will be available during later phases of TEPP. Other available Texas modeling capabilities will be identified during the early operation of TEPP.

Adoption of a Common Computer Language for General Systems Simulation

Available computer languages should be studied during the early operation to determine which is best suited to TEPP. A general systems simulation approach involving several individuals calls for a computer language which is both easily used and available at several locations in the state. Transferability of the modeling system is also a consideration in the choice of a language.

The Definition of Model Components

A general systems simulation approach properly conceived allows the addition and deletion of model components. The definition of model components for TEPP should be completed for the initial two year effort during the first months of the project.

The following components are the likely beginning set: (1) U.S. Macro Component, (2) Texas Macro Component, (3) U.S. and Texas Population Component, (4) U.S. and Texas Energy Demand Component, (5) U.S. and

Texas Oil and Gas Supply Component, (6) U.S. and Texas Coal Supply Component, (7) U.S. and Texas Electric Utility Component, (8) U.S. and Texas Refinery and Petrochemical Component, (9) U.S. and Texas Agricultural Component, (10) Texas Environmental Impact Component, and (11) New Energy Technology Component. This list, including available models, should be updated during the early part of TEPP.

Execution of Staged Model Development

Once the component structure is defined, several stages to full development should be followed. First, available data bases should be surveyed and arranged; missing data should be identified. Second, a skeletal structure of the system should be structured using judgment for specified relationships and parameters to obtain a working version of the modeling system, identifying all relevant input and output variables, policy parameters, and key indicator variables consistent with the Executive Committee's values and problem definitions.

Third, work should be done to provide simplified versions of many of the U.S./Texas components since it is not feasible, nor advisable operationally, to operate these models in the original form as one unit. Contracts should be written early in TEPP to provide this simplification.

Fourth, key components should be developed first, adding the others as time and funding allow. Testing, verification, and validation exercises should be completed at various stages of development for review by the Legislative Budget Board. First priorities should be to build the Texas macro economic model with linkages to the DRI macro economic model, while simultaneously modeling the effects of policy on regional capacity location for key industries on the supply side and improvements in the modeling of oil, gas, and coal supply.

Review of DOE Administrator's Annual Report and National Energy Plans

A current version of the DOE Midterm Energy Forecasting Model should be maintained at Texas A&M for use by TEPP. An evaluation of key parts of the Administrator's Annual Report to Congress and National Energy Plans should be completed by the participants in TEPP. Arrangements for review of separable parts of the report should be arranged by contract where needed. This exercise should be useful as influence and input to DOE's policy work and also as a check on TEPP's view of the importance of various policy proposals.

FOOTNOTES
CHAPTER 3

[9] Greenberger, Martin, Matthew A. Crenson and Brian L. Crissey, *Models in the Policy Process: Public Decision Making in the Computer Era,* Russell Sage Foundation, New York, 1976, page 326.

[10] Johnson, et al, *Agricultural Sector Planning: A General System Simulation Approach,* edited by George E. Rossmiller. Agricultural Sector Analysis and Simulation Projects, Department of Agricultural Economics, Michigan State University, East Lansing, Michigan, 1978, pp. 27-51.

[11] Greenberger, Crenson, and Crissey, Op. cit., p. 24-26.

CHAPTER 4

NATIONAL ADVISORY BOARD REPORT

by Members of the National Advisory Board

INTRODUCTION

The increasing complexity of national (and world) problems together with the increasing capability for construction of large-scale computer models make inevitable the growth in the use of these models by government for forecasting, planning, and decision-making.

In a democratic government there is a special obligation to provide for external communication and verification. Black-box models simply will not do. For outputs of a model used by government to be credible there must be provision for independent analysis of its inputs and inner workings. Moreover, despite its sophistication, the state of the art is, to put it mildly, imperfect. It is, therefore, particularly important that a lively dialogue be sustained between experts inside and outside the government.

The Texas National Energy Modeling Project (TNEMP) was established with three main objectives:

1. to provide an independent evaluation of the Energy Information Administration's (EIA) Midrange Energy Forecasting System (MEFS);
2. to provide recommendations to the Texas Energy Advisory Council (TEAC) concerning the feasibility and desirability of maintaining a national level model for evaluating the impacts on Texas of national energy policies; and
3. to clarify the role of modeling for public policy purposes.

These objectives are discussed in more detail in the TNEMP Executive Summary.

One facet of TNEMP was the establishment of a National Advisory Board (NAB) charged with the following tasks:

1. assessment of TNEMP, its organization and effectiveness in carrying out the project objectives;
2. identification and assessment of substantial issues relative to the structure and operation of MEFS;
3. formulating recommendations for continuation of or follow-up to TNEMP; and
4. statement of other recommendations or conclusions.

This NAB report is organized according to these tasks. The NAB sees its role as vouching for the general integrity of TNEMP, offering procedural advice, providing general reactions to the reports of the eleven Analysis Projects (but not vouching for their detailed accuracy), and formulating recommendations.

AN ASSESSMENT OF TNEMP

The assessment of TNEMP by the NAB included reviewing the organization and its activities, and evaluating the effectiveness of the activities in accomplishing the purposes of the project.

The TNEMP organization has as its working arms the TEAC staff and eleven analysis teams at four Texas universities. Each of these teams was staffed with researchers with already established experience and expertise in their assigned research areas.

The TEAC Executive Director, Dr. Milton L. Holloway, has given substantial technical and administrative guidance to the analysis teams and has provided constructive support to the NAB.

This organization has proven to be most effective in getting the purposes of the project accomplished in a limited amount of time and with limited expense. In contrast to the time and money it would take to build the organization from the beginning, the recognition of the existing resources and a plan for bringing them to bear on this project was insightful and accomplished at lower cost to the taxpayers of Texas.

The NAB considers the full credit should be given to the Executive Director for these insights and for his administrative capability in carrying out the activities of TNEMP. In addition he should be complimented on his effectiveness in dealing with EIA, which led to the penetration of the federal information monopoly and subsequent transfer of MEFS to Texas A&M University.

This penetration and transfer represents a major landmark in modeling technology, as well as a milestone in communication of information concerning national energy policy. The transfer was brought about through the sustained negotiations by the Executive Director with EIA and the technical skills, computer operations, and persistence of interaction with EIA by Dr. Jim Moore of Texas A&M University. He was assisted by the sharp mind and enthusiasm of Mr. John Stone at the University of Houston.

The model transfer is not only important as a landmark advance in modeling technology; it has provided TEAC with the capability for (1) independent verification of DOE results, (2) determining the sensitivity of DOE output to changes in input parameter values, (3) following the consequences of modifications in underlying normative assumptions, (4) better assessment of the limitations of MEFS, and (5) better identification of requirements for measuring regional (Texas) impacts in its own model development.

It is the opinion of the NAB that the successful penetration, transfer, and operation of MEFS represents the most significant contribution of TNEMP. Given the state of the art in computer modeling technology and operations, "transfer" means much more than mere conveyance of a computer tape. It includes conveying an understanding of the model's assumptions, equations, and input data.

In the case of large and complex computer models, "transfer" also requires direct communication between the producing and the receiving modelers concerning the many judgmental factors which are an essential part of model construction but which are not transparent from study of computer tapes and written documentation alone. In principle, these factors should be part of model documentation, but in practice the art of model documentation has seriously lagged behind that of model building.

It is also the opinion of the NAB that TNEMP has made a major contribution to the philosophy of model assessment. The work plan of TEAC was organized around the criteria which a model should meet in its assessment or evaluation. Each analysis team was charged with documenting its efforts to test the models and their components against the four criteria. Whereas some criteria were not applicable to all studies, the concerted, coordinated effort organized around these themes demonstrates the value of this philosophy and its use in model assessment. The underlying philosophy of the four criteria is explored in the report by Johnson and Brown, commissioned by TEAC late in the project. The report raises a number of basic philosophical issues which should be examined and further developed by TEAC and EIA.

Finally, the NAB commends TNEMP for identifying some major modeling questions and weaknesses, as well as bringing to the awareness of policymakers and the public certain energy issues which current policies do not handle adequately or ignore altogether. These questions and issues are discussed in the next section.

SUBSTANTIVE ISSUES RELATIVE TO THE STRUCTURE AND OPERATION OF MEFS

The studies of the eleven TEAC analysis teams have succeeded in identifying several major issues. NAB feels that there are a number of deficiencies of MEFS at the present time. Some of these can be corrected in the near future—some by TNEMP. On the other hand, as in the case of any technology, there is a state of the art in modeling, especially large-scale modeling, which marks the boundary between what is known reasonably well and what is relatively unknown. The user of large-scale energy models needs to recognize this boundary and to realize that sometimes the model will produce results (e.g., forecasts) that are wide of the mark. As was mentioned earlier, the user should not take MEFS as a "black box," but should recognize both its strengths and weaknesses. Above all, the user should realize that one of the most important values of models like MEFS is that they can raise important questions for further study and judgment.

The NAB feels that on balance, MEFS is soundly conceived as a current state of the art modeling system. MEFS has been constructed under severe time pressure and organizational instability. Thus, it would not have been surprising if the consequent hurried decisions had led to more minor flaws than so far seem to be present. It is the belief of the NAB that the current EIA documentation program will repair a high proportion of the minor flaws as well as correcting or ameliorating some of the major ones discussed in this report.

The NAB urges that EIA give careful consideration to the following issues as it continues its development of MEFS.

1. The finding rates calculated in the oil and gas supply models are projected, on the average, to be 20-40% higher than those believed appropriate by Team 2. What is needed is a major reworking of the finding rate functions so as to separate the geological facts from the economic costs to extract. Historically, these have been used together to define what was meant by reserve. The specification of the expected finding rate as a function of exploratory and developmental effort greatly affects the projected *level* of production in the EIA oil and gas supply models (Study 5, Stone & Thompson, University of Houston).
2. The 8% discount rate used in the oil and gas supply models is much smaller than the 15% that the analysis of Study 3 (Rose, Texas A&M) finds needed to motivate investment in oil and gas production. This study also questions the failure of the model to incorporate risk and documents the failure of the model to incorporate important determinants of investment behavior observable in the industry.
3. The oil and gas supply models do not take into account the joint product relationship between oil and gas supplies; the models contain extensive specifications based on the judgments of the modelers (Study 5, Stone & Thompson, University of Houston).
4. The Data Resources, Inc. (DRI) model used by MEFS is highly aggregated and is not linked to states or regions in a meaningful way. The connecting links between the energy market model and the DRI macromodel are grossly inadequate to measure the macro impacts of major structural and regional shifts in investment resulting from energy policy changes. MEFS' demand and DRI models contain too many offline adjustments based on judgments of the modeler, especially in the areas of conservation and regional earning shares (Study 11, Jonish & Taylor, Texas Tech).
5. The ordinary measures of statistical validity of MEFS demand model parameters or sensitivity of the output values are not presented, and indications of the reliability of the parameters can only be inferred by comparison with other study results; such neglect raises questions

about the standards of professional integrity of the modelers and whether standards will be raised in the future (Study 10, Lee & Davis, Texas Tech). The reliability of these paremeters is crucial to energy policy analysis, since they measure both the extent of overall energy consumption levels to energy price changes and the ability of the economy to make fuel substitutions.

6. International effects of energy on the U.S. economy are not dealt with (e.g., the balance of trade, availability of oil imports, balance of payments, and strength of the dollar) and therefore limit the usefulness of the model in answering some of the most important questions of U.S. energy policy (Study 11, Jonish & Taylor, Texas Tech).

RECOMMENDATIONS TO TEAC

TNEMP has thus far been a successful enterprise in two separate senses. First, its substantive objectives have been, by and large, achieved: MEFS has been transferred and put to work in Texas; and eleven analytical studies have been written which raise some significant questions which could lead to an improvement of MEFS. Second, by mobilizing native talents and scholars already engaged on energy problems, TNEMP achieved its initial purposes over a relatively short period of time and at reasonable cost. The NAB recommends to TEAC that the work of TNEMP be continued without delay. The following paragraphs relate to recommendations for methods of continuation and topics for special study.

Recommendation 1

That TEAC maintain a national-level energy model. Continued maintenance of a national-level model will enable Texas government and industry to get timely answers to "what if" questions related to economic and non-economic impacts of changes in national energy policy. MEFS provides a useful starting point for analyses of regional energy and economic development issues which are coupled to national policies and/or which require interregional considerations. TEAC should concentrate on improving MEFS both through recommendations to EIA and by integrating new research results into its own version of the model. Toward this end, TEAC should tend toward concentrating its resources on research and modeling at the state and regional level with attention to integrating these results with MEFS and perhaps other national models which can accommodate regional detail.

Recommendation 2

That TEAC continue its cooperation with Texas universities. Continued cooperation with Texas universities will permit TEAC to profit from substantial expertise on a "pay as you go" basis and to change the size and scope of its studies more easily than if a staff of state employees is established and maintained. This principle applies to location of the computer model as well as that of analysis studies. The central TEAC needs to be large enough to provide overall management, direction and an interface between state-level users and the university study teams.

Recommendation 3

That TEAC organize and support an annual assessment of U.S. energy conditions and policy, and of special developments and policy issues relating to Texas. TNEMP has demonstrated the capability of TEAC, the supporting institutions, and the various research groups in the Texas university system to penetrate the EIA modeling system, and to provide a critical review and analysis of the capability of that system to support research on particular energy policy issues. TNEMP assessment has been conditioned by attention to policy issues of special interest to Texas and to the other energy producing states. Various TEAC recommendations, which this Advisory Board endorses, propose to continue this type of activity and to coordinate more carefully research and modeling relating to energy policy and regional economic development.

Toward that end, this Advisory Board believes that TEAC should consider organizing and sponsoring an annual review of energy policy issues. Such a review would provide a forum for summarizing and presenting the policy implications of new research of TEAC, the supporting institutions, and the universities, as well as providing an alternative perspective to federal analyses. We recommend that the TEAC Annual Energy Policy Review (AEPR) follow publication of the EIA's Annual Administrator's Report, providing a counter analysis, where appropriate, to the EIA presentation. The AEPR should provide a useful "focusing event" for those members of the energy policy research community in Texas who have professional contributions to make in both the national and regional discussions. If combined with an effort to annually transfer the MEFS to Texas A&M, the TEAC AEPR would provide an opportunity to maintain the continuing assessment of MEFS would contribute to the national policy debate, and would present counter analyses to the nation from a Texas perspective. The ability of Texas to provide such counter analyses within the same modeling framework as the EIA will serve to improve greatly the quality and credibility of the TEAC AEPR and the analyses of EIA.

Recommendation 4

That TEAC lead the way in regional analyses. Texas, as a large and highly involved state, should lead the way in undertaking solutions to regional problems related to MEFS, working from the bottom up.

Recommendation 5

That TEAC continue its analysis of oil and gas stripper wells. MEFS does not seem likely to analyze oil and gas stripper wells in any detail. Because of increasing importance of these sources, a careful study of their role in energy supply seems highly appropriate as part of continued TNEMP operations. Few other parts of energy production have such extensive and direct contact with citizens of Texas. Even if oil and gas prices are deregulated, it will be useful to monitor the productive response of this modest but not insignificant sector of the U.S. energy system. There will be a transition period when analysis of stripper wells may be useful in protesting the interests of Texas producers.

GENERAL COMMENTS AND RECOMMENDATIONS

1. The Department of Energy (DOE)/EIA procedures (1) for validating and verifying models, (2) for documenting and transferring models, and (3) for handling model outputs which lead to policy analyses and evaluations, all need continuing review and examination by "disinterested" committees of analysts, by users, and by representatives of the general public. Several external sources for aid in achieving these goals are:
   A. National Academy of Science/National Research Council
   B. National Science Foundation
   C. Bureau of Standards
   D. establishment of an Office of Model Assessment reporting to the U.S. Congress
   E. by contract with professional societies (e.g., Operations Research Society of America and/or The Institute of Management Sciences).
2. We recommend that EIA extend its plans to annually transfer MEFS to the Argonne National Laboratory Software Center to include the participation of TNEMP to ensure the maintenance of a "living model" at a non-EIA site.

   The Energy Conservation and Production Act (ECPA) of 1976 provides evidence of congressional concern regarding the validity of the EIA models. Expressing these concerns, Congress (1) established the Professional Audit Review Team (PART) to provide an annual review of EIA activities including modeling, model applications, and data development; (2) established a requirement for EIA documentation of PIES (now MEFS); and (3) required EIA to develop procedures providing public access to the EIA models. PART produced its first report to Congress in December, 1977 and is nearing completion of its second report. Also, a major effort at documentation of MEFS was initiated, although not with complete success as noted in the various technical reports of TNEMP's study. On the third point, public access, EIA has formulated a plan to each year transfer that version of MEFS used in developing the Annual Administrator's Report (AAR) to the Argonne National Laboratory Software Center.

This Advisory Board has been impressed with EIA's efforts to meet both the letter and the spirit of the various legislative mandates of ECPA-76. Further, the various plans described to us by Dr. Elizabeth MacRae concerning future documentation and assessment of new model developments seem reasonable and consistent with good scientific and management practice. We are less sanguine about EIA plans to satisfy the public access requirements of the legislation. In observing the EIA/TNEMP interactions, it has become increasingly clear to us that the procedures for physical transfer of MEFS from EIA to Argonne will satisfy only the letter and not the spirit of the congressional concern for public access to EIA models. The fact that TNEMP will be continued, and that others are expressing interest in executing their own assessments and analyses with MEFS suggests that in the future EIA will continue to have many requests for assistance in MEFS transfer. We recognize that EIA has many demands upon its resources, and must balance its requirements to perform analyses and studies for the Department of Energy (DOE), the Administration, and the Congress, against requests from non-government agencies for assistance in obtaining and applying the model. Accordingly, we recommend that EIA seriously consider modifying its approach to providing public access to the model to include TNEMP as a partner in developing the procedures for ensuring that a "living model" can in fact be transferred. The precise details of such a plan would be worked out by TNEMP, EIA, and Argonne, but key elements might include developing procedures to ensure that replication by TNEMP of important analysis results reported in the AAR be ensured, and "certification" by TNEMP that the documentation and information supporting the model transfer to Argonne is sufficient to ensure that the transferred model is not "sterile," and can be used by organizations with the necessary resources and capabilities.

3. EIA should consider expanding or rotating its list of contractors. Regional balance tends to counteract some of the regional political bias that is to be expected from most contractors. Phasing in new contractors also avoids "inbreeding" and enhances the chances for correction or errors and for innovations.

4. We urge that the DOE/EIA ensure that a comprehensive response to TNEMP assessment is provided, both an overview response to be included with TNEMP's report, and a more comprehensive and substantive report to be published separately.

   An important feature of TNEMP's process has been the involvement of an EIA liaison, Dr. Elizabeth MacRae, to assist in coordinating model transfer, in providing information about and interpretation of, EIA model development activities, and in responding informally to the critical comments of TNEMP assessment teams. This Advisory Board feels strongly that Dr. MacRae is to be highly commended for her contribution to TNEMP's assessment process, and for her general professionalism and competence. From our perspective, however, many of the issues raised by TNEMP assessment groups have not yet been adequately addressed by EIA. We want to ensure that these points are properly evaluated and reflected in future EIA modeling and research plans. It would be surprising if all the TNEMP critical comments were accepted by EIA. The nature of model development in support of policy research is such that many judgments and compromises on allocation of the resources, on approach, and on interpretation are required. TNEMP has provided EIA the opportunity to include a response with the TNEMP report in order to ensure that the EIA position and perspectives are communicated together with the critical assessment. However, timing of publication may prohibit inclusion of the fullest appropriate EIA response, and so this Advisory Board recommends that EIA consider preparing a comprehensive response.
5. EIA's response to TNEMP has been positive. As a new agency making a fresh start, it would be wise for EIA to sustain this process of interaction in the future and wise for external analysts to sustain their monitoring of models, their assumptions, and outcomes. Indeed, unless this is done, it seems doubtful that the federal government will ever gain any credibility in its models and in policy recommendations based on outputs from these models.
6. In the short term, the State of Texas (through TEAC) needs to adopt its own version of Recommendation 1 above and also be involved in 2.
7. TEAC should alert U.S. senators and representatives from Texas to its activities and encourage them to draw up legislation as appropriate to fulfill the recommendation of this study to make use of TNEMP's modeling capabilities in the development of energy policy and to evaluate energy policies proposed by others.

CHAPTER 5

DEPARTMENT OF ENERGY REVIEW COMMENTS

by Elizabeth Chase MacRae and Frederic H. Murphy[12]

OVERVIEW

At present, there are many individuals and organizations active in the area of energy modeling and analysis. The Energy Information Administration (EIA) within DOE is just one such organization, and, although it is somewhat larger than most other modeling groups, is not particularly distinctive in terms of its data, methodology, or staff. What is unusual, or even unique, is that EIA carries out its modeling activities under close public scrutiny.

The sensitivity of energy issues, about which there are widely divergent opinions, has led EIA to adopt a number of strong policies and procedures with respect to its modeling and analysis so as to avoid the fact or appearance of bias in its forecasts or analyses. In preparing forecasts, EIA policy is to present a range of projections to emphasize the inherent uncertainty of any statements about the future. All analyses, whether initiated by EIA or requested by others outside of EIA, are carefully documented and are made publicly available. In addition, an extensive program of model documentation has been initiated which, over the next few years, will create and maintain useable information on all operating models for a variety of audiences. Finally, a regular cycle of model review and assessment is underway to provide insights for both analysts and decision-makers as to the strengths and weaknesses of EIA models and to guide the continuing programs of model development.

Although EIA reacted initially to the idea of a TNEMP review of EIA models with some skepticism, the professional qualifications of the individual analysts and the guidance from the National Advisory Board has made TNEMP a valuable adjunct to EIA's ongoing program of review and assessment.

The benefits of DOE, however, were not what might be expected. Very little new was learned about the models from the TNEMP reviews, and many of the weaknesses identified have already been eliminated or are being addressed through EIA's program of model development. In some cases, the reviews reflected a lack of familiarity with the state-of-the-art in energy modeling and with the availability and quality of energy-related data. Moreover, virtually all of the suggested improvements, feasible or not, were in the direction of more detailed models, in contrast to requests by DOE policy makers for simpler models.

Although TNEMP did not provide EIA with significant new insights into the models themselves, it did provide several rather more subtle benefits. First, it made very clear to EIA what were the difficulties and costs in successfully transferring operational models to entirely new users, even when those users devote substantial effort themselves to the transfer as was the case with TNEMP. Even though this was a pilot project and thus understandably involved a certain amount of inefficiency due to EIA's learning process, it still represented a considerable drain on EIA's resources.

Second, it was encouraging to find that many of the problem areas found by independent analysts were the same as those we had already identified ourselves, even though we did not necessarily agree upon the solutions.

A number of general observations about model use and review are in order. To begin with, it is important to clear up a common misconception about the relative importance of models in the activities of government. Responsible decision-makers do not often, and should not, rely heavily on models. Although results from MEFS are often part of the background material on which policies affecting the mid-term energy future are based, it is naive to suppose that DOE decision-makers put their faith in models to any great extent. The proper role of models, or any sort of analysis, is to sharpen the intuition of the decision-makers by keeping track of a variety of inter-related factors. If model results are not intuitively clear to the decision-makers, and cannot be made so, then the model is of no value and its results should not form the basis for decisions. For a decision-maker to use model results which are not intuitively explainable is inexcusable.

A second misconception is that models should ideally embody the concept of scientific objectivity, free of judgmental factors. While it is certainly essential to identify and document the judgment which enters into both the development and use of a model, it is impossible to purge subjective factors without destroying the relevance of the model to issues of importance. Mechanical development and operation of models, untempered by judgment, rarely produces informative results. And, as was mentioned above, the purpose of models is not so much to provide the answers as to improve the intuition and judgment of the decision-maker.

A useful common structure was provided for the TNEMP reviews through the establishment of the four evaluation critiera: workability, clarity, validity, and verifiability. However, as might be expected from a pilot effort involving many different analysts, the application of the review criteria varied across models. In addition, the review of models, like their initial development, typically involves a good deal of judgment even when explicit criteria are specified. In the TNEMP exercise, the four criteria specified were not wholly objective either in their definition or in their application.

The most objective of the criteria, verifiability, deals primarily with the question of whether the model, as implemented on the computer, actually behaves the way the modeler intended. In very simple terms, verifiability involves checking for typographical errors in the computer program and ensuring that the program is correctly written. In this area, TNEMP analysts provided a useful service by identifying several "bugs" in the models which had not been noticed by EIA.

Clarity, the next most objective of the evaluation criteria, refers to how easily the structure of the model can be described in intuitive terms. Some judgment on the part of the reviewer enters into the question of clarity since there can obviously be differences of opinion with respect to what is intuitive or clear. Nonetheless, a key consideration in any measure of model clarity must be the adequacy and quality of model documentation. Although the current quality of documentation is comparable to that in other government agencies, EIA agrees with the TNEMP conclusion that current quality is not sufficient for an agency which must make its models accessible to the public, and which must scrupulously maintain an aura of objectivity. Moreover, lack of adequate model documentation means that training of new analysts within EIA is done somewhat inefficiently through reliance on memory and word-of-mouth. EIA's current program of model documentation is aimed at creating and updating documentation for all operating models, written in various forms for audiences ranging from the interested layman or user through the technical researcher. Since EIA has some 70 different models, most of which are not part of MEFS, upgrading all the documentation will necessarily take a few years.

Quite a bit further down in order of objectivity is the criterion of validity. The definition used by TNEMP measures the validity of a model by comparing its structure to accepted theory in the areas of economics, geology, and engineering. Unfortunately, there is very little consensus, especially in the critical area of economics, on theories which are sufficiently well quantified to form a basis of modeling.

Models are simply a way of organizing available information and judgment so as to illuminate particular issues. In most cases, a model attempts to describe real-world relationships which are not precisely known, and which are based on a myriad of decisions and choices by consumers, producers, and government. Thus, there is an inherent uncertainty about the nature and form of the relationships in a model about which there will always be disagreement.

Models in the energy area are meant to describe the aggregate or average behavior of a wide variety of decision-makers, including for example consumers who choose what type of automobile to purchase and how much to drive it, energy-using businesses which determine their investment in plant and equipment, energy-producers who decide how much exploration to undertake for new oil, electric utilities which plan for future capacity expansion, and various government organizations which establish regulations and legislation. No one really knows how these consumers, businesses, or governments react to various economic, demographic, regulatory, or attitudinal factors.

Although there are a few generally agreed upon concepts which characterize the behavior of these decision-makers, there is certainly no agreement on the quantitative details of any of the relationships. Thus, there will always be an element of judgment in models and always some disagreement among equally competent individuals as to what the appropriate relationships are, and consequently, different measures of model validity.

In addition to underlying theory, there are two other factors which determine the choice of model structure and thereby impact upon the measure of validity. The first is availability of data with which to quantify the theory. Even if there is agreement regarding the approporiate structure of a model, lack of relevant data to give substance to that form means that some less satisfactory formulation must be constructed.

The second factor is the level of disaggregation needed in a model for addressing the issues. Even though theory may describe the detailed workings of some aspect of the energy system, it is inefficient to burden the model with the extra complexity if a more aggregate representation is informative enough for analysis of the relevant issues.

The fourth evaluation criterion, workability, is perhaps the least objective of them all since it deals with the applicability of models to certain energy issues, whether or not the models were ever intended to address those issues. The workability criterion might more reasonably be applied to the full model development program where it could indicate where some issues were not being covered in EIA's development plans. Thus workability, as defined by TNEMP, is not really an objective characteristic of a particular model.

A more relevant characteristic would be whether the model is capable of sharpening the intuition of the user. In the case of the MEFS models, there have indeed been situations where the application of a model to a particular issue produced unexpected insights for the decision-maker. Unfortunately, this feature is not really susceptible to independent measurement outside of the context of actual model use for a decision-maker.

All things considered, EIA is pleased with the results of TNEMP, particularly with the successful transfer of several interrelated models to independent users and a different computer. Without the considerable efforts of both EIA and TNEMP staff, the project could easily have become a frustrating exercise with no useful results.

SUMMARY OF RESPONSES

The Texas Energy Advisory Council (TEAC) review of the submodels of the Midrange Energy Forecasting System is in a long traditon of reviews of Energy Information Administration models, including reviews by the National Science Foundation and Resources for the Future. This review differs from previous ones, because of the interests of the reviewing organization. TEAC is constituted to comment on the analysis of energy issues, in relation to the Texas position, in legislative debates and is not a disinterested scientific body, as are the aforementioned groups. Accordingly, the TEAC goals and emphasis are different from those of other reviewers.

In previous cases there was no formal EIA response. Instead, there was an extensive reliance on dialogue among the model designers and commentators. As direct contact was not extensive in this case, TEAC has provided an opportunity for EIA to include a formal response within its report on the Texas National Energy Modeling Project. The organization of this response to their review is as follows. First, the response regards comments on the entire system. Next, there is a short response to comments on the individual submodels. Finally, the comments by the analysts knowledgeable about the individual details are contained in Part II along with the relevant critique.

General Comments on the TEAC Report

The TEAC reviews of the submodels of the Midrange Energy Forecasting System range from excellent and informative to barely competent. Individual comments vary from insightful to polemic. Except for some specific instances, the bulk of the misunderstanding comes from inadequate documentation of the models or the time lag between the documentation and model improvements. For example, in reference to statements about model improvement in the 1974 *Project Independence Report (PIR)*, it is said that the "resolve . . . to mend the deficiencies (of the 1974 version of the model) was lost." Yet, since 1974, many important changes have taken place. These include:

- A new demand model with many coefficients regionally estimated (in contrast with the ad hoc 4 by 10 rule for scaling elasticities in the *PIR*).
- An expanded utilities model.
- Variable drilling rates and ramped price paths, which were incorporated into the oil and gas supply model.
- Submodels which have been added, showing the effects of regulation on oil and natural gas.
- The National Coal Model which was developed to provide the coal supply curves.

- Three different refining models which have been tried, and a fourth which is in the final stages of development.

What has not changed is the overall architecture of modeling system interactions and the overall format (as opposed to substance) of the integrating model. In terms of format, the integrating model even has the same basic naming conventions of the Bonner and Moore ECCM model, the predecessor to the Project Independence Evaluation System (PIES). As the authors of the TEAC report do not know the full history of the Mid-range Energy Forecasting System (MEFS), it is understandable that they would not know what changes have taken place. This reflects the documentation problem inherent in a rapidly changing modeling system. For example, although 17 volumes of documentation and 12 technical appendices were produced as part of the 1977 *National Energy Outlook*, EIA has not spent the extra time and money to extend the software for MEFS from a modeling system, which is a useful aid to analysts within EIA, to a modeling system product, which can be easily utilized by anyone. To illustrate this difference, it costs a lot more to write a commercial software package for linear regression, than to code a few, homegrown routines for internal use.

Whether MEFS can be converted into a modeling system product is unclear. This is because the individual models of the system are loosely knit. Essentially, MEFS provides a mechanism which presents individual judgments and analyses in a coherent package. The most important TEAC finding, and probably the most frustrating to TEAC as an outside group, is:

> "MEFS' results are highly dependent upon the diffuse judgments of many modelers within DOE. This means that outsiders must necessarily have considerable faith in the integrity and judgment of the modelers themselves, as well as the 'objective scientific' performance of a computerized modeling system."

The only fault we can find with this statement is that by terming the people involved as "modelers," there is the implication that their expertise is modeling at the expense of understanding energy. The people making the judgments, as opposed to cranking out numbers, are, first and foremost energy analysts. We have been routinely accused of being model builders and model runners, as opposed to analysts. The TEAC experience has shown that one of our main contentions is true, that is, human judgment, which at least is making sure model results make economic sense, is the dominant influence on EIA forecasts.

The main difference between a modeling system and a modeling system product, beyond the detail in documentation, is the extent to which the system must be foolproofed to protect the unsophisticated user. An example is a problem that arose in TEAC's developing some new scenarios different from the EIA scenarios. An unbounded solution in the integrating model occurred due to abnormally altered oil and gas supply curves. This happened because TEAC was substantially modifying the finding rates in the oil and gas supply model. TEAC described this event as "a new problem never encountered by DOE implying the need for more verification tests." The truth of the matter is that problems of a similar nature continually occur in a modeling system that has not been fool-proofed. For example, a problem that has not happened in TEAC runs, but periodically occurs here, is nonconvergence of the integrating model, because of interactions of the average-cost-pricing mechanism for electric utilities with the demand model reapproximations. Without changing the equilibrium forecast, this problem is handled by reversing a few steps in the program logic or altering some smoothing coefficients to break a cycle. As this is a relatively infrequent event, it is not worth the cost of the extra work to find a permanent solution.

The diagnosis of the TEAC problem by EIA staffers was completely routine. The problem of a new scenario specification is one of adapting a modeling system to the specification. Solving minor technical problems in matrix generation, however, is only the first step in making sure the results make sense. For TEAC to use MEFS for its own analyses, it will have to build its own team of energy analysts who understand the models, rather than rely on EIA analysts for debugging its scenerios. EIA cannot continue to provide support to resolve computer system inconsistencies or other implementation problems, nor should it be expected to interpret TEAC results.

In light of the scenario modification problem, there is the following statement: "Verification tests illustrating the consistency of the model's behavior with underlying logic and expectations of the modelers are virtually nonexistent." This sounds like poor management by EIA staff, but does this statement have any meaning? Does it mean that supervisors have failed to personally audit all code written by employees? In this case, the statement is true, but would anyone expect this? Does it mean that model results were not compared across different scenario specifications to see if the differences, as well as the levels, are reasonable? If this is what is meant, the statement is false and unwarranted. The whole reason for TEAC's interest in MEFS is that the system is used for forecasting and policy analysis. By definition, this means that many different scenarios with a wide range of assumptions are used. The Annual Reports to Congress contain a series of forecasts that illustrate the uncertainties within the current legislative environment. Subsequent analyses for clients reflect the effects of policy differences over the uncertainties highlighted in the Annual Report. A major portion of the time devoted to producing a report is spent on making sure results match economic theory and sound judgment. This process invariably leads to questions, more model runs and new scenarios.

Before commenting on the specific submodel reviews, three more points should be made. It is stated that the National Energy Plan analysis is not verifiable. This is not completely true. The most contentious piece of the analysis has been reproduced by the staff that did the original work. The integrating model runs have been redone replicating the results of 2 years ago. What cannot be reproduced are the outputs of the models that provide inputs to the integrating model.

There is the statement that the model is overly simplistic in certain areas. This is the standard comment which is usually combined, as it is in this case, with complaints about modeling system complexity. EIA has tried to maintain a balance between necessary detail and the advantages of easily interpreted relationships in model emphasis to avoid a weak link. What is oversimplification to one reviewer is unnecessary detail to another. Operational considerations, balanced with the need for detail to represent policies, are the issues that must be resolved in determining the total amount of detailed structure allowed in MEFS.

In its recommendations, TEAC suggests that EIA follow a policy of "Rotating its list of contractors. Regional balance tends to counteract some of the regional political bias that is to be expected from most contractors." EIA staff make an effort to hire the most capable and cost-effective contractors available. For example, firms such as Bonner and Moore, PACE, and Mortada, all Texas firms, have worked for EIA. We have contracted geologists from Louisiana and Colorado. It also should be noted that the person who headed the original National Energy Plan analysis at the FEA as well as the head of the integration division during that period had both been professors at Texas universities prior to joining FEA. The local contractors regularly retain experts on energy industries from around the country. We have always endeavored to retain a high degree of objectivity, even if our forecasts may not have been used objectively.

In the absence of any evidence supporting the presumption of political bias on the part of any of our contracts, we intend to continue our policy of trying to hire the very best, without discriminating on the basis of geographic origin.

Comments on Individual Chapters of Part II

There is no disagreement with the content of Chapter 1, Part II, and Jim Moore did an admirable job in moving the MEFS system to the Texas A&M computer. It is important to recognize, however, that EIA is not constituted to be a commercial, software house, and, thus, cannot continue to devote a significant amount of staff time to aiding others in reconstructing our models on systems incompatible with ours. Also, it would take a substantial amount of the taxpayers' money to generalize our system to represent all possible scenarios that non-EIA users may want to specify. Our goal is to make sure we properly represent EIA scenarios and make them reproducible by others.

Half of the assessment reports focus on the oil and gas supply model. They cover the methodology for incorporating U.S. Geological Survey (USGS) resource estimates and the estimating of finding rates, the representation of investment behavior, and the behavior of the computer model.

The next chapter raises one of the most important issues associated with oil and gas supply modeling. What is the significance of the dramatic fall-off of oil and gas finding rates, that is, the average number of barrels per foot drilled? This problem is discussed adequately in the book. We have long been concerned with this problem ourselves and the problem has been studied by our contractors. The report, "Analysis of Drilling Productivity in the Lower-49 States," prepared in January 1979, deals with this issue. The conclusion of this report is that the low productivity is due to the radical, post embargo, price shift. This led to the drilling of low risk, low productivity, but profitable, wells. There has been a lot of in-field drilling, for example, which shows up in a sample of Texas wells that were examined.

EIA is continuing to monitor the data and will remain on top of the issue. In sensitivity runs, in the future, we may adjust the initial finding rate with the same resource base to show the consequences of the different possibilities.

In the chapter dealing with the investment process, several questions about the representation of investment behavior are raised. The first question is whether cash flow is a better determinant of drilling behavior than the net present value (NPV) of a project. There are two aspects to this question. The first is whether NPV is the better indicator of what ultimately would be drilled. If we assume that no project that is expected to be unprofitable is undertaken, and we assume that every opportunity that is expected to be profitable is eventually undertaken, then we have to agree that the expected NPV is the best measure of what would ultimately be produced, and this is how the NPV calculations are used.

The second question is what is the major determinant of the timing of drilling. The drilling heuristics within the oil and gas model are not based on any economic theory and are one of the weakest aspects of the oil and gas model. The timing issue, however, is not as important as one may initially think. The bulk of the additions to proved reserves will come from the outer-continental shelf (OCS) and Alaska. The major deteriminant of drilling in these areas is the leasing schedule. A relatively smaller portion of the new production will come from onshore. Production from yet to be proved reserves will represent only a portion of the production in the midterm.

Cash flow is probably the least important deteriminant of drilling behavior. The industry has developed many non-financial arrangements to deal with the problem, such as, farm-outs and joint ventures. Probably the most important determinant of drilling rates is the value of the information gained from wells drilled.

The other questions raised relate to what is a reasonable rate of return to use for the NPV calculations. The issues here are what operators think they need and what is a reasonable risk premium. The reviewer believes an 8 percent real rate of return is too low and it should be around 15 percent. The 15 percent comes from an operator survey by the reviewer. Considering the number of financiers that do not know the difference between real and nominal interest rates, one must be careful in interpreting the answers by operators. Also, it is not clear to what extent a high interest rate is used internally as a counter to innate optimism on the part of geologists. Finally, because of overhead, the interest rate used to evaluate a single project has to be higher than average, if a corporation is to achieve an average rate of return. Historically, oil companies have achieved an average rate of return; and, according to the reviewer himself, the risk in the oil and gas business has fallen since the embargo, reducing the need for a risk premium on the rate of return.

The chapter dealing with stripper wells treats a relatively small portion of the oil supply. There is no explicit representation of stripper wells in the model, but developments in the area are monitored and analyzed.

The chapter dealing with the behavior of the oil and gas supply model, shows a lot of hard work and thought on the part of the reviewers. Many valid criticisms about the model are raised. There has not been much effort expended directly on the model, other than data updates, nor has enough been provided on model documentation. The reason for this is that there is an ongoing effort to develop replacement models. For instance, a new model for Alaskan hydrocarbons was delivered in time for the 1978 ARC.

A new OCS model is in the final stages of completion. Both this and the above model operate at the individual opportunity level. The OCS model has the added feature of directly incorporating the USGS Conservation Division data by prospect. The Conservation Division geologists are taking a major role in the data analysis, eliminating organizational barriers to understanding. A new onshore model is under contract for development. Its specification is nowhere near complete and the comments from the TEAC review of the current oil and gas model will influence its design.

In terms of the current model, two of the points that have been raised need to be addressed. The reviewers correctly state that not only is the level of oil and gas supply of interest, but also the supply elasticity is important. In their experience with the model, as the finding rate is reduced by 50 percent, the supply elasticity for new gas increases from .2 to 2.32. The important point is that this is new gas and not total gas. With the lower finding rate, the new gas supply is much lower. Therefore, even with the higher elasticity on new gas, the elasticity of total supply remains very low in the midterm.

The reviewers state that because of Btu equivalent pricing of coproducts, gas supply estimates are biased downward and oil supply estimates are biased upward. This fails to take account of the role of the integrating model. Each supply step has a minimum-acceptable, supply price equal to the value of the bundle of products. The integrating model requires that the cost of the bundle equal the weighted values of the primary product and coproducts. The bias described can occur only if the supply curve steps are not sufficiently dense, a problem we try to avoid. The Btu-equivalent pricing of coproducts does, however, bias the regional distribution of drilling away from being equal on a marginal profit basis. Nevertheless, this is a minor problem.

The reviewers of the coal supply model exercised the National Coal Model, by varying some of the key assumptions such as mine life. However, because of the size of the coal resource base, there was little alteration in the coal supply curves in the range of the quantity demanded within the equilibrium framework. Although this specific case of extending the mine life has not been run by EIA, this compares with the many different scenarios we have run in the normal course of business. Coal supply and prices are the most stable component of the energy system.

The concern was expressed that the coal supply curves "indicate a more rapid, industry production response than observed empirically." This comment illustrates why the focus on energy should be on the role of markets. The slow growth in demand has been due to market forces and government regulations affecting the effective cost of coal, and reducing the growth in consumption. For example, there currently is overcapacity in coal markets due to the lower than expected growth in demand.

The review of the refinery model says that the model is too simple and inflexible to be useful for a wide range of scenarios. The easiest way to describe the refinery model is that it gives the historic product yield by crude type, bounds the relative product prices, and accounts for Btu's consumed during refining. This is a true placeholder model as described by Vanston and Baughman in Chapter 8. The question becomes one of how the model is calibrated. Currently, the model is initially calibrated by using the Bonner and Moore RPMS, a system for detailed representation of refineries. Whenever there are substantial changes in the refinery operating environment, such as higher prices or the need for a dramatically different product slate, the model is recalibrated. The recalibration is fully automated and done regularly.

To reduce the need for adjusting a placeholder model, we are working on our fourth attempt at a different refinery model. We are engaged in an attempt to scientifically reduce the size of a large scale, refinery model, minimizing the product price and quantity distortions with each successive revision.

The reviewers express the concern that all refining would take place in the regions with the highest historical yield. Their concern would be true if the refinery model were exercised outside of the full integrating framework, which it is not. Transportation costs, plus the effect of the activities that set the relative product prices, create an environment where this does not occur.

The comments on synthetics in this chapter are parallel to the refinery comments. Beyond government subsidized products, synthetics will contribute very little to midterm energy supply. Since the midterm has been extended to 1995 in the 1978 ARC, more detail, such as individual representations of the different processes, has been added.

The reviewers in the chapter about electric utility sectors essentially said, that the electric utilities model is a "placeholder" model, that is, a model that is a vehicle for transmitting preordained conclusions. The reviewers describe the model as the simplest possible representation of electric utilities. Unlike the refinery model, the electric utilities model cannot be dismissed as a placeholder. The model has very tight constraints on the type of new capacity that may be added. However, the constraints, except for 1995 nuclear capacity restrictions, reflect the real world, lead times in building certain types of equipment. Using the fuel prices of the year of interest for determining capital equipment selection may be a crude theory of behavior. However, in the midrange, there is, in reality, little flexibility beyond current construction plans for electric utilities.

Because of the period of interest, the model contains more than the simplest possible representation of dispatching. The model distinguishes between existing and new capacity, it does not have a predetermined merit order, and has extensive detail on hydropower dispatching. The Baughman-Joskow model, in comparison, does not distinguish between existing and new capacity, operating characteristics and has a prespecified merit order.

The comments on the transportation model are an expanded version of the standard comments one would made about a corporate plant location, distribution model. In establishing transportation costs, one resorts either to simple averages or to complex, mixed-integer, programming models. There is little intermediate ground with improved predictive capability, outside of expanded regionalization. At the same time, expanding the number of regions greatly increases the model size. For oil products, transportation costs form a relatively small portion of total costs, making expanded detail unnecessary. For natural gas, the region size issue is overwhelmed by the economies of scale question for pipelines (which is in need of further study). For coal, transportation costs form a significant portion of the total delivered costs. Because of this fact, in the integrating model there is an expanded transshipment network and the consumption of coal is shared out to subregions using population. Contrary to what is said in Chapter 2, Part I, there are no serious coal transportation constraints in the model.

In terms of costs of transportation links, the reviewers expressed the concern that future rail rates are underestimated, because the current rates do not reflect fully allocated costs, as defined by the Interstate Commerce Commission (ICC). However, the costs used in MEFS are based on fully allocated costs, not current tariffs, since it was expected that rail rates would rise from their historical low level once coal shipments expanded, with it becoming a sellers' market from the railroads' point of view.

In the review of the demand model the authors ask the standard question economists would ask about the demand model as an entity in itself. They show an adequate understanding of the demand model, but do not understand how the demand model relates to the rest of MEFS. The chapter as now constituted is a great improvement over the previous drafts, but still displays an apparent lack of understanding of how a hierarchical modeling system works.

As an example, the authors of the chapter say

> "price perturbations employed in converting the dynamic econometric model results into a static representation of demand . . . may have led to incorrect specification of the static demand functions because of the quantity dependence on lagged quantity, which in turn is a function of lagged price."

A long run, demand elasticity is estimated by incorporating a set of price paths into the demand model. This simulates the year-by-year price response over the time period of interest. One then takes the different demand levels in the given year of interest, using the price in that year, and calculates arc elasticity. The result, of course, is implicitly dependent on the price path. All supply and demand curves used in the integrating model are implicitly dynamic, being derived from satellite models that are explicit in their dynamic structure. The role of the equilibrium model is to tie together, in a given year, the results from these dynamic models.

The authors do not understand how the demand model interfaces with the supply structure. The problem is one of distinguishing the difference between a solution procedure and the model being solved. The role of the step function approximation is as an algorithmic device that has no influence on the final solution, outside of convergence tolerances. The step function approximation is refined between successive linear programming solutions, until the prices and quantities from the linear program are within tolerance of the demand model. There are extensive descriptions of the algorithm in all the documentation and in technical journals.

As the summary chapters were written using the previous draft of the chapter on the demand model, the summary chapters contain serious misunderstandings. The full TEAC report suffers as a consequence. The kind of obvious error that remains in the summary, but not the chapter, is "some variables such as heating degree days were dropped . . . before projections were made." It should be clear that because the output of the demand model represents a summary of the consequences of the input assumptions, one need not pass all demand model input information along with the demand model output to the integrating model.

Drawing on the chapter on the demand model, Part I includes the following statement: (the procedures used to estimate the demand model) . . . "raises questions about the standards of professional integrity of the modelers." This comment alludes to the lack of summary statistics on estimated coefficients. The reviewers also suggested using simultaneous estimation approaches. In fact, the lack of summary statistics and lack of a simultaneous estimation approach was due to a simple problem. Commercially available estimation software at the time was inadequate for the job. The only available package capable of handling the large amount of data incorrectly calculated the summary statistics, and there was no software available for simultaneous estimation. To understand the difficulty of estimating even the most rudimentary demand model, one need only look at the arbitrary rules used to determine the natural gas elasticities in the PIR. The state of software for estimation has improved and such statistics will be produced in the future.

The authors make a point that the model results change only slightly with major changes in the elasticities. This is because the starting prices for equilibration are chosen close to the final solution to speed convergence. Consequently, arbitrarily changing the elasticities, while using the same starting price quantity pair, has little influence on the final solution because the final prices are close to the starting prices. The case where elasticities were changed along with oil and gas finding rates would not show many differences beyond increasing shortages and fuel substitution, because interstate gas is average-cost priced and is mostly old gas under price regulation; and unmet demand is determined by alternative fuel prices. The authors should run a high oil import price case to see how demand would change.

The essence of the last chapter is that although the models used are state-of-the-art, they are inadequate. The macroeconomic-energy interaction needs to be placed in perspective. Because energy is a small portion of GNP, macroeconomic changes have a greater impact on energy than changes in energy markets have on the macroeconomy. The role of macroeconomics in influencing energy demand has always been recognized, but the methodology for quantifying the effects of energy policies on macroeconomics is only just developing. Several attempts to capture this interaction have been made in previous years, and, for the current, annual report, new procedures have been developed. As the forecasts are extended further into the future, higher energy prices will have more of an influence on the macroeconomy. To deal with the problem, Hudson and Jorgenson are working on an expanded version of their model to help us understand these issues. We are looking at the alternative approaches as well. Improved understanding of energy economy interactions is one of our highest priorities.

FOOTNOTES
CHAPTER 5

[12]Elizabeth Chase MacRae is currently Assistant Administrator for Program Development, Energy Information Administration. During most of the TNEMP exercise Dr. MacRae was Deputy Assistant Administrator for Applied Analysis, Energy Information Administration, and served as the EIA liaison to TNEMP. The overview portion of the Department's Review Comments was written by Dr. MacRae. Dr. Frederic Murphy is an Operations Research Analyst, Energy Information Administration, and wrote the Summary of Responses to the 11 studies of Part II.

CHAPTER 6

WORKSHOP ON SUBSTANTIVE MODELING ISSUES[13]

WASHINGTON, D.C.

27 August 1979

by Milton L. Holloway

Members of the Texas National Energy Modeling Project (TNEMP) and the Energy Information Administration (EIA) met in the Forrestal Building in Washington, D.C. on August 27, 1979 to discuss substantive modeling issues raised during TNEMP's review of EIA's Midterm Energy Forecasting System (MEFS). The meeting was structured to provide overview statements by EIA's Assistant Administrator for Applied Analysis, the Executive Director of TEAC, the Chairman of the National Advisory Board for TNEMP, and EIA's liaison to TNEMP in the first part of the meeting; this was followed by individual workshop sessions for (1) oil and gas supply models, (2) coal supply and electric power models, (3) demand and macro models, and (4) MEFS operation and computer model transfer and, finally, ending the workshop by reports from the four groups with discussion by all participants.

OVERVIEW STATEMENTS

Roger Glassey, Assistant Administrator for Applied Analysis, began with an opening statement focusing on EIA interaction with TNEMP. He said that the transfer and operation of the models by TNEMP had demonstrated several things of benefit to EIA in understanding the requirements for successful transfer. First, the exercise highlighted inadequacies in existing EIA model documentation. Second, the exercise helped clarify what documentation and procedures are necessary for model access. Third, the exercise made it clear that the mailing of computer tapes and written documents is not adequate to accomplish the kind of transfer needed for new users to be able to set up and operate large scale models such as MEFS.

Glassey said there are two views of models. One view is that models should be part of a scientific process having the usual attributes of reproduceability, transferability, etc. If modeling is an art, this view is, of necessity, wrong. The other view is that models are "aids" or "extensions" of analysts, a part of a process to allow an analyst to understand problems and to sharpen his intuition. In the TNEMP exercise EIA was transferring only a part of the models since all of the knowledge of the analyst cannot be transferred even with close cooperation. The part that depends on the personal input of the analyst was not entirely transferred.

Milton Holloway, Executive Director of TEAC, spoke next and focused on (1) a review of why TEAC/TNEMP got involved in the evaluation and (2) some philosophical issues that underlie the TNEMP exercise. He pointed out that the early interest by TEAC and others in Texas was in looking at the natural gas model because of the issue of natural gas price deregulation, and at the entire modeling system because they did not believe that President Carter's policy proposals emphasizing conservation and taxes over production incentives would achieve the import reduction goals shown in the National Energy Plan (NEP). They simply wanted to have access to and review the models used to justify the NEP.

Holloway then reviewed the definition of objectivity and its relationship to the four evaluation criteria as used by TNEMP. He contrasted this idea of objectivity to that implied in Elizabeth MacRae's overview response to the TNEMP report; that is, according to MacRae, objectivity is primarily related to the absence of political influence on the analysis or analysts. An objective analyst, as explained by Holloway, means one who is willing to have the concepts produced by his investigation and research undergo scrutiny of the criteria of workability, clarity, coherence, and correspondence, and to abide by the results. Objective analysis is work which has undergone and passed such tests.

The four criteria definitions as reviewed by Holloway are:

—*Workability*. A model is workable if it serves some practical function, identifies issues, suggests a resolution of issues, or in general produces results that relate to the practical problems of the user.

—*Clarity*. A model is clear if its behavior and results can be translated into terms understandable to other modelers, laymen, and policymakers.

—*Verifiability* (correspondence). A model has verifiability characteristics if its results agree reasonably well with reality.

—*Validity* (coherence). A model has validity characteristics if it exhibits the logical consequences of accepted theories; if it does not, then the theories or model require modification. To state it differently, a model has validity characteristics if its behavior corresponds to the modeler's expectation as per its design; the model must do what the modeler says it will do; the model must behave in a manner consistent with the underlying logic or theory upon which it is built. Coherence is at least partially dependent upon the understanding of the disciplines best represented by the current body of theory.

Holloway expressed the opinion that Congress was wrong in setting up EIA as an independent branch of the Department of Energy. EIA is not producing "objective" forecasts and analyses, in the sense explained above, simply because it is "independent" of the policymaking functions of the administration. Interaction with interested parties — those who have some stake in the issues — is essential to the development of objective work. He said that TNEMP has done a great deal to move MEFS a step closer to objectivity; it still has a way to go.

Robert Thrall, Chairman of the National Advisory Board to TNEMP, summarized the National Advisory Board's role in and view of the TNEMP evaluation exercise. He explained that the membership was made up of people nationwide and that their primary role in the effort was to vouch for the integrity of the process of evaluation rather than to judge the specific critiques in the eleven individual reports. He said the Board could indeed say that a major good-faith effort was made on the part of both TNEMP and EIA to achieve the transfer of the large modeling system and to achieve interaction between the modelers and model evaluators.

Thrall further reviewed the major recommendations of the National Advisory Board which are:

1. TNEMP can be helpful in EIA fulfillment of the legislative mandate to make its modeling available to the public; EIA should fund a continuing transfer of a "live" version of MEFS. The Texas A&M center is clearly an appropriate agency for such transfer.
2. By legislative mandate, TEAC has been assigned the task of making an annual assessment of EIA's Administrator's Annual Report to Congress; DOE/EIA should cooperate by making modeling, data, and other information available for the timely completion of this task.
3. An on-going dialogue should be maintained between EIA/PE and users (such as TNEMP/TEAC) on modeling and policy work. Such a dialogue could utilize the congressionally authorized Regional Advisory Groups.
4. EIA should respond substantively to the findings and recommendations of the TNEMP report. If this response is timely it could be included in the published TNEMP report.

According to Thrall, a major problem of EIA is achieving credibility of their models. Everything that increases credibility is a plus; the ability to transfer the MEFS model to Texas and to do some sensitivity analysis of the major model components has and will continue to add credibility.

Thrall also emphasized the need for EIA to provide some regional balance in contract work for model development and policy analysis; there is inherently a regional bias (not deliberate) because of the problems individuals are accustomed to dealing with and the limited perspectives of individual analysts. A rotation of contractors will help balance this kind of influence and help prevent the repeat of past mistakes. Longer advertisement of competitive contracts is needed.

Elizabeth MacRae, Assistant Administrator for Program Development who served as EIA's liaison to TNEMP, summarized her view of the TNEMP exercise. She said that early in TEAC contacts, during debate over the National Energy Act and while DOE was being organized, DOE appeared initially to be uncooperative in responding to TEAC's request for background materials. When the Freedom of Information Act request was filed, it was misdirected to Policy and Evaluation. They lacked sufficient interest, given the status of the national debate and the problems of organizing the Department of Energy, to treat the request seriously. By the time the request reached EIA, the massive personnel changes and departures delayed EIA's ability to respond. Once there was some organizational stability, EIA met the TEAC request, and in addition went beyond the requirements of the Freedom of Information Act by having EIA personnel help to resolve inconsistencies between the computer system used by TEAC and that used by EIA.

MacRae said that there are still disagreements between EIA and TEAC concerning what the models do or what they should be used for, but the attitude on both sides is now much more constructive.

MacRae commented on the four criteria used by TNEMP in the evaluation process. A major misconception in the TNEMP report is that models (EIA models) embody scientific objectivity. Models are primarily tools for sharpening one's intuition. She said that of the four TNEMP criteria, clarity and verification were reasonably objective criteria but that workability and validation were subjective in their application. It is difficult to test models relative to accepted theory (validation) since there is much disagreement on the correct theory. Further, it is difficult to evaluate models concerning their relevancy to practical problems since they (models) are usually somewhat general and interpretations are needed to apply results to specific problems. Furthermore, models do not really address issues; only analysts address issues.

MacRae finished by saying she enjoyed the interaction with the Texas group and was pleased that together we had been able to effectively transfer MEFS to Texas.

WORKSHOP SUMMARIES

After lunch the workshop participants reported back to the full group, listing specific issues discussed, agreements and disagreements on the issues and implications for future research and model development.

Electric Utilities and Coal Supply

The electric utility and coal supply group reported first. Since the Texas evaluator of the coal supply model was not able to attend the meeting, the coal supply model was not discussed. One specific agreement as reported by Martin Baughman was reached on the electric utility model:

1. The discussions resulted in a better understanding of the electric utility model representation. Distinctions must be drawn between two types of behavior represented in the model—capacity expansion behavior versus production behavior. The module *can be* constrained to represent many different behavioral theories within the Midterm Energy Market Model, though only one—that of cost minimizing behavior—has been employed in published EIA analyses. While the modeling of electricity production is adequate, however, expansion behavior neglects problems of cash flow and financing; thus they may simulate expansion behavior which is inconsistent with the industry financing capacity. There is no financing feedback check in the model; EIA must assume that industry expansion can be financed. It was stated that EIA has recently issued an RFP for model development to characterize electric utility financing, regulation, and cash flow considerations on the capital expansion behavior of the industry.

Some issues remained unresolved; the workshop terminated with the following disagreement:

1. TNEMP reviewers argue that there were other weaknesses of the model which could have been pointed out; for example, the model does not represent the interregional electrical energy interchange that might be possible under certain conditions, nor are maintenance schedules checked for consistency with regional economic dispatch calculations. EIA argues that capacity is accounted for by Federal region served. Where necessary fuel shipping charges reflect nine month costs, i.e. to simulate interregional transfers. Interregional transfers are not directly represented due to convergence problems with average cost pricing.

Oil and Gas Supply

The oil and gas supply group reported next. Agreements were reached on the following issues:

1. Concerning directionality (directing drilling specifically toward oil or gas) it was agreed that although the EIA model does exhibit some directionality when interfaced with the integrating model, there is a methodological problem which needs attention. Empirical work and better data than presently exist will be required before the problem

can be adequately modeled. It was agreed that to the extent drillers do not know whether they will find oil or gas with a prospect, the model does not capture the increase in oil supply in response to higher gas prices and vice versa.

2. On the matter of finding rate methodology and estimates, it was agreed that the EIA model exhibits a higher finding rate for the current time period and the near term future than is evident from data since 1974, because data from years prior to the embargo were included when estimating finding rates by EIA methodology. More importantly it was recognized that in order to resolve the issue with a high level of confidence, major studies are needed to better characterize the resource base, including pool-size and distribution characteristics. Such studies will provide empirical evidence to test whether reserve additions from new drilling behaves in a relatively smooth declining fashion over time or in plateaus or "epochs." Current data are inadequate for answering this question. A major concern is that drilling productivity data are not structured to relate drilling activity to resultant reserve additions. It was agreed that in order to improve oil and gas supply modeling, resource characterization must receive high research priority; public sector agencies must rely on USGS work, such as Circular 725, and this needs improvement and update.
3. Concerning the lack of representing oil and gas as joint products, it was agreed that interface with the integrating model yields results which simulate joint products, under most scenarios. Under some scenario specifications (e.g., in case of rapidly declining world oil prices), however, the correct results would not be obtained from the model. Further, the present means of representing joint products is not the best methodological way to deal with the problem. It was agreed that this methodological problem must be addressed in the new models being developed, and EIA reported that it is to the extent data are available.
4. Further empirical and theoretical work is needed to adequately reflect the decision-making process of the industry.

The identification of a number of recognized weaknesses in the current model has led EIA to contract for the development of a new model now under way.

Several issues remained unresolved; the workshop terminated with the following disagreements:

1. On the matter of directionality in drilling, EIA states that the oil and gas supply models represent perfect directionality. This is because the supply models are used to generate separate supply

curves for oil and gas. Higher prices for gas means a greater supply of gas is available with the attendant increase in drilling. Oil supply is not increased by higher gas prices except to the extent that the higher price associated increases the value of the total product. The TNEMP reviewers argue that the EIA model does not represent perfect directionality; the modeling procedures do not accomplish it. Perfect directionality implies that drilling activity (and consequent finds) are dynamically re-allocated over time in response to changes in the relative profitability of oil and gas products. Changes in profitability occur both because of different price paths and because of different rates of decline in the finding curves. The EIA model reallocates only the additions to drilling stock in each year. Existing capacity (whenever brought on line) is not reallocated. This is not perfect directionality. A more directional drilling allocation would probably yield more gas relative to oil. EIA states the allocation of existing rigs problem is not significant as long as the price paths chosen lead to additions of rigs for both fuels.

The more fundamental issue is an empirical question of the extent to which drilling really is directional; i.e., how well can drilling be focused on one or the other of oil and gas. It is well understood that resolving this question requires the collection of much better data than is currently available.

2. On the matter of bundle pricing, the problem is one of consistency between the pricing assumptions in the oil and gas models and the pricing results of the Integrating Model. The TNEMP reviewers insist that pricing is only an approximate solution to the problem. The EIA statement is that any combination of shadowprices which produce a zero reduced cost for a supply activity in the Integrating Model would produce a non-negative net present value if input into the supply models. TNEMP reviewers argue that this is not necessarily true because the revenue components for oil, gas, and liquids in the supply models are weighted by different factors for royalties and taxes. These weights are effectively assumed equal in the Integrating Model because they are not directly represented. Thus, consistency is not guaranteed and the quantitative significance of the pricing discrepancy depends upon: (1) the magnitudes of the differences in royalty and tax factors, (2) the magnitudes of the differences between shadow prices and input prices, and (3) the closeness of net present value to zero in the oil and gas supply model. This is a methodological problem in the interface between component models and integrating model. A straightforward, though perhaps computationally difficult, solution offered by TNEMP reviewers is

establishing another cycle of iterations between shadowprices in the integrating model and input prices in the component supply models. EIA agrees that the different royalties introduce a source of error. However, the royalty represents a small portion of the production cost, with the difference in royalty rates a fraction of the royalty rates; and co-products are a fraction of the value of the primary product.

3. On the matter of empirical evidence concerning the behavior of the finding rate, there was disagreement concerning recent data. EIA representatives argue that it is not clear what changes have taken place since the embargo in the types of drilling undertaken throughout the country. The acceleration in activity could explain the drop-off in finding rates. The TNEMP reviewer argues conversely that Texas data clearly show no significant change in the drilling mix in the post embargo period. In fact, a slight increase in high-risk drilling has occurred.
4. The appropriate rate of discount for the oil and gas exploration and development industry was left unresolved. The TNEMP reviewers argue that the TNEMP studies based on survey results as well as other evidence imply the need for a higher discount rate than used by EIA. A higher risk is inherent in the oil and gas production portion of the industry; therefore, comparison of survey results with the average rate of return for the industry as a whole from historical data is inappropriate. EIA representatives argued that the historical evidence is that the nominal rate of return in the oil and gas industry has been 9.7 percent and therefore, the real rate is lower than the 8 percent used in the model; the rate of return for this industry is not greatly different from other industries. As 9.7 percent is the average rate of return, the rate of return on the marginal projects, i.e. the ones for which the rate of return is important, is much lower. EIA argues further that the way the TNEMP survey question was posed was misleading. The typical executive questioned is unlikely to understand the difference between real and nominal interest rates even when it is explained to him. Also, higher internal rates of return may just be used to counter optimistic geologists. It is also unclear, EIA argues, that oil and gas exploration and development are more risky than, say, refining, given current regulations. The TNEMP reviewers point out that they were very explicit on the distinction between real and nominal rates of discount and that personal follow-up questioning made it clear that individuals answering the questionnaire did indeed recognize the difference.
5. The influence of capital rationing and lack of agreement of the

model with capital budgeting theory by the industry was not agreed upon. The TNEMP reviewers argued that capital rationing is, in fact, practiced in the industry and that proper theoretical representation requires representation of the demand for capital and the marginal capital supply function faced by the industry. Net cash flow directly affects the marginal cost of capital and therefore capital investment; these factors are not captured in the EIA model. Questions of the validity of interpretation of the Texas survey results were raised by EIA. They pointed out that capital rationing practiced by individual firms does not translate into capital shortages in the overall market, as rationing is usually a reflection of the cost of capital to the firm.

Various points of discussion in the general meeting concerned the oil and gas models.

Macroeconomic Interface

Several agreements were reached concerning the adequacy of the energy/macroeconomic interface. A full dicussion of the important issues raised in the report was not possible, however, since neither of the authors of the report nor the reviewer, were present at the workshop. Some issues were resolved as being differences in interpretation or emphasis regarding the TNEMP report. Also, the TNEMP reviewers dealt with the DRI model version utilized in the AAR-77, not the DRI 1979-C version which corrects some problems. Other issues of agreement included:

1. The MEFS/DRI macro model interface is not a fully integrated, regionalized modeling system and therefore does not provide adequate information for understanding regional macroeconomic impacts of energy policy changes. The DRI/MEFS system does not account for the regional investment and disinvestment decisions of industry important to an understanding of policy impacts. Ad hoc studies of, for example, industry relocation impacts must be done by EIA to provide such information as things are now structured. At the same time, it is not clear that a fully integrated regional modeling approach is available. Some current work by EIA with the Hudson-Jorgenson model may be helpful in better modeling the investment process at the national level. Recognizing the inadequacy of the regional detail, EIA said it is their policy to use only national numbers in analyses, the regional detail is used to provide one level of disagregation below the forecasts.
2. Iterating between the DRI model and MEFS is inadequate as a means of obtaining simultaneous solutions for the two models. EIA has apparently made some improvements in achieving consistent

macro-energy forecasts in their work with the 1978 AAR version than was the case with the 1977 AAR version evaluated by TNEMP. Important methodological and operational difficulties are inherent in interfacing a large, quarterly, partly recursive, disequilibrium model with a small, annual, equilibrium model. The weak linkages between the two models result in the loss of significant information.

3. Measurement of world trade effects of energy policy changes is not adequately reflected in the DRI model, affecting the model's ability to capture the balance of trade impacts of such factors as a change in the world oil price. Such analyses must now be done "off-line" and imposed on the model solution. It is not clear, however, that data currently allows a better approach.
4. It was agreed that empirical evidence of apparent complementarity of energy and capital *in the aggregate* for the economy is in fact tentative.

Some disagreements were left unresolved:

1. TNEMP reviewers emphasized the great difficulties that external users face when attempting to work with models, such as DRI, which have been considerably "managed." The reference is to the large amount of DRI add-factoring and exogenous variables influencing the model results and behavior. The problem is not necessarily whether add-factoring is appropriate but whether add-factoring makes it difficult for outsiders to guess the unknown quantity/quality of such adjustments in generating model results, and further that the contribution of add-factoring is not published with results and made available to the user. EIA personnel believe this is less serious that the TNEMP report implies. If one simply accepts the DRI model as an acceptable state-of-the-art model of the U.S. economy then the important question is how outputs from MEFS are treated in DRI. A recent study by DOE has examined the extent of change in the DRI forecast from a change in MEFS price paths—the major linkage between the models.
2. Disagreement remained about the usefulness of inflation and employment impact estimates from energy policy changes as measured by DRI/MEFS. Some believe that other economic policy or business cycle influences so overwhelm energy that the numbers generated from energy market and policy changes are not reliable especially when a range of macro-economic forecasts are used. Others believe the impact numbers, as a marginal change to whatever else is going on in the economy, are reliable for energy policy analyses. TNEMP reviewers believe that the linkages are so weak,

especially given the unknown influence of add-factoring, making the results very suspect.

Demand Model

Most disagreements seemed to be resolved concerning issues raised in the TNEMP report on the demand model. Several important agreements were reached:

1. Adequate documentation does not exist explaining procedures for others to implement their own scenarios, performing the kind of elasticity sensitivity tests which were performed by TNEMP. EIA has not completed such sensitivity tests either and therefore could not provide good guidance to TNEMP in performing the sensitivity analyses. It seems clear, however, that the exogenously specified price paths so influce and "drive" the demand model, and that sensitivity tests cannot be performed involving MEMM without changes in the price paths. EIA (at the workshop) could not define a clear procedure for testing the importance of reliable price elasticity parameters on the overall MEFS solutions.
2. The current demand model does not adequately deal with the dynamics of the change in capital stock and its influence on the end-use of energy due to the nature of econometric models which rely on the range of historical relative prices and their effects. Because of this problem, and the inadequacy of data, EIA is replacing the industrial demand model and is collecting extensive new data on the end-use of energy.

Remaining disagreements involved methodology:

1. TNEMP reviewers argued that a simultaneous equation approach is superior to the reduced form equation approach used by EIA for the demand model estimation. However, EIA did not agree with this conclusion.
2. TNEMP reviewers argue that experiments by EIA are needed to clarify the issues of the importance of econometric demand parameters on the overall model solution. One can interpret a set of static demand curves from the model, such as those constructed by TNEMP, but sensitivity test involving these static demand curves do not provide much information since time lag structures dominate the behavior of the model. TNEMP is premature in concluding that the demand model is price insensitive. EIA views the demand curves as a one year slice of the dynamic evolution of energy demand, where the price paths are an integral part of the analysis.

Model Operation and Access

Several agreements were reached concerning the transfer of the computer code and data bases and the model operation in Texas:

1. Problems of putting the modeling system up at Texas A&M were relatively minor but nontrivial given the size and complexity of the system. The problems were mostly software related, and the process should be substantially easier a second time.
2. EIA is willing to entertain an unsolicited proposal to maintain a "live" version of the model in Texas. Such a project, in addition to other benefits, would provide EIA with a back-up system.
3. The process of putting the modeling system up on another computer and performing experiments as was done in the TNEMP experience adds credibility and forms a base for improved communications between two analytical groups. Access by outsiders improves public acceptance of the modeling results.

General

The matter of contract advertisement by EIA for TNEMP members bidding on modeling and policy work was discussed. Most contracts are done on a competitive bidding basis and handled by individual project managers within EIA. No common solicitation list exists. Timing for response is typically two to three weeks after issuing an RFP. EIA does produce a contract proposer's list of intended work upcoming, issued prior to the actual RFP. A copy of the current year's work program was promised to be sent to TNEMP members in early October.

It was agreed that EIA would put together an internal solicitation list and that TNEMP should be put on the mailing lists for both the advance program notices and RFPs when issued.

Roger Glassey closed the meeting by a follow-up response to the opening remarks of the morning. He said that EIA independence from policy has many benefits. Since EIA does not design policy it can objectively review and provide analyses of such work. Working with the policy staff is a mutually educational experience. EIA must explain and prove its numbers for policymakers to accept them.

Regarding continuing transfers of MEFS, Glassey said that it is desirable for EIA to have copies of MEFS used by various groups. Perhaps there are too many national models being used and built; we might all be better off improving existing models.

---

[13]While the contents of this chapter were drafted from notes taken by both EIA and TNEMP participants and subsequently reviewed by both parties, final responsibility for the interpretations and content of the report as recorded here must rest with the author. Not all issues were adequately addressed in the limited time available and, further, some key members were unable to attend and therefore the workshop results did not benefit from their perspectives and tend to under-emphasize issues in their area.

APPENDIX A

DOCUMENTATION OF USES OF MEFS

1. Federal Energy Administration, *Project Independence Report,* November, 1974.

   This publication was designed to evaluate domestic energy policy alternatives (natural gas pricing, strategic petroleum reserve, research and development, and conservation programs), "not only in terms of our vulnerability to supply disruptions, but many other important factors, such as economic and social impacts, environmental effects and regional differences." The levels of energy supply, demand and imports; the economic cost of an oil import interruption; air, water and solid waste pollution; balance of payments; and inflation and GNP growth rates were analyzed with respect to the policy alternatives.

2. Federal Energy Administration, *National Energy Outlook 1976,* February, 1976.

   This publication attempted to "sharpen the discussion of policy alternatives" by explicitly evaluating "the effects of alternative price regulation policies, environmental controls and shifts in the energy resource mix." This was done by analyzing the effects of the following policy options on capital requirements and energy production, price and import levels.

   a. Aggressive Domestic Energy Development — Accelerated federal OCS leasing, increased investment tax credit, increased Alaskan Pipeline capacity and eventual price decontrol.

   b. Energy Conservation Actions — Auto efficiency standards, van pooling, thermal efficiency standards, appliance efficiency standards, accelerated industrial energy conservation, improved airline load factors, electric utility load management, and elimination of gas pilot lights.

   c. Price Regulation and Controls — Domestic oil and gas prices of $7.50/barrel and $1.15/mcf, respectively, and $9/barrel and $1.25/mcf, respectively, in 1975 dollars.

   d. Accelerated Coal-Nuclear Electricity Production.

   e. Restricted Supply — A moratorium on yet-to-be permitted nuclear power plant construction, decelerated leasing of the OCS through 1980, restrictions on mining and buring of coal including heavier reclamation costs and severance taxes, and mandatory use of scrubbers.

3. Federal Energy Administration, *1977 National Energy Outlook,* unpublished draft, January 15, 1977.

   This report is very similar to the 1976 version. "This NEO reports on progress in developing a better understanding of the complexities of the energy situation and provides an examination of the processes which influence future energy alternatives. This year's NEO:

—Explicitly relates the international energy supply and demand environment to the energy suituation and the energy problems facing the United States;

—Reports improved analysis of the factors influencing oil and gas supply; and

—Presents an explicit analysis of the effects of the regulation of the wellhead price of gas."

4. Executive Office of the President, *National Energy Plan Supply, Demand and Pricing Estimates: A Background Paper,* June 27, 1977.

The provisions of the NEP, namely the oil and gas pricing, electric utility, conservation and tax policies, were analyzed as follows:

a. "The PIES Model was used to produce a base case forecast of quantities and prices to provide a context for analysis."

b. Each provision of the NEP was analyzed using the major PIES modules in conjunction with off-line "models of passenger car fuel demand, energy demand in buildings, coal conversion, and dispersed solar and geothermal supply." The effects of the NEP on energy consumption, price and import levels for major fuel types consumed in each sector of the economy were projected to 1985.

5. Federal Energy Administration, *PIES Forecasts of the 1985 Impacts of National Energy Legislation: The House Passed National Energy Act and the Administration's Proposal to the Senate,* September 19, 1977.

"The impacts of the House Bill and Senate proposal on projected 1985 energy consumption and prices were analyzed using the Federal Energy Administration's Project Independence Evaluation System (PIES)." Projections of 1985 energy savings and prices in each sector of the economy for both proposals were compared to the base case projection.

6. Energy Information Administration, *Annual Report to Congress 1977,* April 1978.

The effects of different levels of demand, geologic accuracy and imported oil prices on energy imports and energy consumption rates and prices of fuel types for each sector of the economy were projected to 1985 to 1990 and compared to the Base Case. This was accomplished "by incorporating the macroeconomic and energy-related assumptions into the Project Independence Evaluation System (PIES)." Further analysis of other energy-related issues was accomplished by linking the following models "directly or indirectly to PIES."

—Data Resources, Inc. model incorporated "PIES projections of energy prices, gasoline consumption and imports . . . together with macroeconomic assumptions used to drive the PIES projections."

—Bureau of Labor Statistics Input-Output model incorporated the macroeconomic projections to determine employment levels in 129 sectors.

—A regional earnings distribution and model projected the level of sectoral earnings by state.

—Regional Emissions Projection System incorporated PIES estimates of regional fuel consumption to estimate future air emissions of particulate matter, sulfur oxides, nitrogen oxides, hydrocarbons, and carbon monoxide for each of the 243 Air Quality Control Regions in the United States.

—Comprehensive Human Resources Data System incorporated PIES projections of residential energy prices and other assumptions to estimate household energy expenditures by socioeconomic group.

7. Energy Information Administration, *Analysis Memorandum No. AM/1A-7802, An Evaluation of Natural Gas Pricing Proposals,* June 14, 1978.

"This paper presents the results of an EIA analysis of several natural gas pricing proposals . . . as compared to continued current regulations on natural gas markets." The PIES model was used to project the following impacts of the various proposals:

—Wellhead prices in the Southwest for new natural gas
—Domestic natural gas production
—Natural gas shortages
—Natural gas prices by sector
—Natural gas consumption by sector
—Energy imports
—Producer revenues.

8. Energy Information Administration, *Analysis Memorandum No. AM/1A-78-17, A Further Evaluation of Natural Gas Pricing Proposals,* August, 1978.

This analysis differs from the previous analysis only in that it assumes "that incremental pricing provisions do not go beyond the pipeline, and that prices of new supplies are 'rolled in'." The previous analysis "used the assumption that the incremental pricing provisions do not go beyond the pipeline."

9. Bureau of Land Management, *Final Environmental Impact Statement Crude Oil Transportation System: Valdez, Alaska to Midland, Texas.*

The Project Independence Evaluation System linked the supply projections for the Alaskan North Slope, Beaufort Sea, and the National Petroleum Reserves (which were estimated through individual engineering-oriented assessments) with demand scenarios for each sector of the economy. This was used to project the effect of alternative imported oil prices on domestic oil demand.

10. Energy Information Administration, *Analysis Memorandum AM/IA/78-18, An Air Emissions Analysis of Energy Projections for the Annual Report to Congress,* September 6, 1978.

    "Fuel use changes by airshed were derived by the conservation and growth adjustment model. This model uses . . . Project Independence Evaluation System (PIES) fuel projections as . . . inputs. It estimates the emissions implications of the various fuel use projections. . . . These adjustments are applied to each AQCR in such a way that the aggregate for all AQCR's in a Federal Region equals the total from PIES."

11. Economic Regulatory Administration, *Draft Programmatic Environmental Impact Statement: Fuel Use Act,* November, 1978.

    This analysis "was designed to forecast the level of increase in the use of coal and alternate new facilities" by region "as a direct result of the fuel conversion regulatory program. The coal production projections," also by region, "used in this analysis were designed to maximize impacts expected to result from the conversion regulatory program. Environmental and health impacts were assessed based on these projections. . . . [T]he analysis depends heavily on the PIES modeling framework."

12. Energy Information Administration, *An Air Emissions Analysis of the Energy Projections for the Annual Report to Congress,* September, 1978, EIAC Order No. 400, Pub. DOE/EIA-0102/16.

    Supplements the EIA information on six series of energy projections for each of the years 1985-1990. (REPS, based on MEFS output)

13. Energy Information Administration, *Alternative World Oil Price Scenarios for NEP-II,* November, 1978, EIAC Order No. 401, Pub. DOE/EIA-0102/19.

    Presents the results of an analysis of 15 alternative world policy scenarios with exogenous variables including: economic growth, Organization of Petroleum Producing Countries (OPEC) production capacity, Communist net oil trade, non-OPEC production, and National Energy Act estimates savings. (OMS, based on IEES and MEFS output)

14. Energy Information Administration, *Analysis of Proposed*

Regulations Implementing the Powerplant and Industrial Fuel Use Act, November, 1978, EIAC Order No. 404, Pub. DOE/EIA-0102/21.

Analyzes impacts, costs, and benefits of alternative regulatory programs that the U.S. Department of Energy could adopt to implement the Powerplant and Industrial Fuel Use Act. [MEFS and CUMIS (a model provided by a non-EIA contractor)]

15. Energy Information Administration, *A Distributional Analysis of the 1985 Energy Projections for the Annual Report to Congress of the Energy Information Administration,* June, 1978, EIAC Order No. 410, Pub. DOE/EIA-0102/25.

    Examines the distributional effects among household groups of the energy events represented in the projections developed for the Energy Information Administration's *Annual Report to Congress, 1977.* (CHRDS based on MEFS output)

16. Energy Information Administration, *An Evaluation of Energy Related Tax and Tax Credit Programs,* July, 1978, EIAC Order No. 413, Pub. DOE/EIA-0102/8.

    Analyzes the impact of the following as associated with the National Energy Act: (1) tax credits for energy conservation and solar investments; (2) petroleum and natural gas user taxes and investment tax credits; and (3) crude oil equalization taxes. (MEFS including MEMM, RDFOR and CUMIS, HIRST, and JACKSON Solar Market Development Model)

17. Energy Information Administration, *An Evaluation of Natural Gas Pricing Proposals,* July, 1978, EIAC Order No. 415, Pub. DOE/EIA-0102/5.

    Analyzes the effects of several natural gas legislative alternatives considered in 1978. (MEFS Oil and Gas Model, CUMIS)

18. Energy Information Administration, *Fact Sheet: Forecasted Automobile Gasoline Consumption,* May, 1978, EIAC Order No. 419, Pub. DOE/EIA-0102/1.

    Describes the methods used by the Energy Information Administration to forecast automobile gasoline consumption for the Energy Information Administration *Annual Report to Congress, 1977.* (MEFS Demand Model, Transportation Component)

19. Energy Information Administration, *Forecasts of Principal Petroleum Product Demands (1979, 1985, 1990) by DOE Region and Four Macroeconomic/Geologic Scenarios,* August, 1978, EIAC Order No. 420, Pub. DOE/EIA-0102/12.

    Gives estimates by Department of Energy Region and four macroeconomic/geologic scenarios for the domestic demand of principal refined petroleum products. (MEFS)

20. Energy Information Administration, *Further Evaluation of Natural Gas Pricing Proposals,* July, 1978, EIAC Order No. 421, Pub. DOE/EIA-0102/10.

Examines natural gas pricing proposals for 1985 based on current regulations and several alternative pricing proposals. Adds to the report, *An Evaluation of Natural Gas Pricing Proposals.* (MEFS, MEMM)

21. Energy Information Administration, *Historic and Forecasted Energy Prices by DOE Region and Fuel Type for Three Macroeconomic Scenarios and One Imported Oil Price Escalation Scenario,* November, 1978, EIAC Order No. 425, Pub. DOE/EIA-0102/27.

Forecasts energy prices at the Department of Energy Regional level through the year 2000, and provides historical fuel prices for 1975 and 1976. (MEFS)

22. Energy Information Administration, *International Energy Supply and Demand Support Documentation for the Administrator's Annual Report, 1977,* August, 1978, EIAC Order No. 428, Pub. DOE/EIA-0102/14.

Documents the analysis associated with Chapter 4, "International Energy Supply and Demand," of the *Annual Report to Congress, 1977,* Volume II. Addresses energy supply, demand, electricity generation, economic growth, and balance of payments. (IEES, OMS using MEMM as input)

23. Energy Information Administration, *Natural Gas Pricing Analysis —Senate Proposal,* April, 1978, EIAC Order No. 436, Pub. DOE/EIA-0102/36.

Compares the cost of a Senate Proposal, dated March 7, 1978, to the natural gas pricing provisions under HR 8444, the House version of the National Energy Act. It is a differential analysis of the incremental revenues transferred to natural gas producers. (MEFS, Oil and Gas Supply Model)

24. Energy Information Administration, *Refined Product and Crude Import Tariff Analysis,* July, 1978, EIAC Order No. 448, Pub. DOE/EIA-0102/31.

Provides estimates of the effects of proposed tariffs on imported refined products and crude oil. (MEMM)

25. Energy Information Administration, *Revenues to Natural Gas Producers Under Alternative Regulatory Assumptions,* June, 1978, EIAC Order No. 450, Pub. DOE/EIA-0102/28.

Provides forecasts of natural gas producer revenues through 1985. Uses three alternative assessments of undiscovered recoverable gas resources. (MEFS Oil and Gas Supply Model)

26. Energy Information Administration, *Medium-Run Oil and Gas Supply Model 1977 Update,* December, 1977, EIAC Order No. 430, Pub. DOE/EIA-0103/13.

Documents the Medium-Run Oil and Gas Supply Model. Describes the data used in the oil and gas submodels and updates the oil and gas economic submodels.

27. Energy Information Administration, *Mid-Term Oil and Gas Supply Modeling System Methodology Description,* November, 1978, EIAC Order No. 431, Pub. DOE/EIA-0103/17.

Documents the Mid-Term Oil and Gas Supply Modeling System. Describes the approach used to develop projections of oil and gas supply for the Lower-48 States and portions of Alaska.

28. Energy Information Administration, *Modeling Natural Gas Regulation in the Project Independence Evaluation System,* July, 1978, EIAC Order No. 432, Pub. DOE/EIA-0103/7.

Describes the architecture of the natural gas submodel in the Project Independence Evaluation System (PIES). Included are an overview of PIES and a description of the supply allocation, demand, and shortfall mechanisms.

29. Energy Information Administration, *National Coal Model—Coal Supply Curves,* September, 1978, EIAC Order No. 435, Pub. DOE/EIA-0103/2.

Discusses the coal supply component of the National Coal Model (NCM) and presents a set of coal supply curves which are the primary NCM outputs.

30. Energy Information Administration, *Oil and Gas Supply Curves for the Administrator's Annual Report,* September, 1978, EIAC Order No. 439, Pub. DOE/EIA-0103/4.

Presents supply curves for crude oil, natural gas, and natural gas liquids for the years 1985 and 1990. These curves are generated from an Oil and Gas Model by the Department of Energy.

APPENDIX B

EVALUATION OF DOE/MEFS TRANSFERABILITY AND DOCUMENTATION

by Jim Moore

Data Processing Center
Texas A&M University

EXECUTIVE SUMMARY

This study deals with the specifics of the MEFS system transfer and operation on the Texas A&M computer. The set of software systems installed and satisfactorily tested include:

1. Integrating Model
2. Regional Demand Forecasting System (RDFOR)
3. RDFOR Preprocessors
   a) Price Preprocessor
   b) Macro Preprocessor
   c) Parameter Preprocessor
   d) Initial Prices Preprocessor
   e) Conservation Shifts Preprocessor
4. Midrange Coal Supply Model
   a) does not include the entire National Coal Model (NCM)
   b) two University of Houston Fortran Modules have replaced the need for the Gamma Matrix Generator and Report Writer
5. Midterm Oil and Gas Supply Model System
6. Standard Integrating Table Preprocessors
   a) Oil Preprocessor
   b) Gas Preprocessor.

The various systems were installed and results checked against DOE/EIA results from the same systems operated on the DOE computer in Washington, D.C. Minor differences in the DOE check results and those produced at Texas A&M have been attributed to the following:

1. Version 1 of MAGEN used at TAMU versus Version 2 of MAGEN used at DOE
2. MPSX/360 used at TAMU versus MPS3 used at DOE
3. Different starting basis
4. Possible different alternate solutions
5. Replacement modules used in the Coal Supply Module.

The check results were exact except for the integrating system and the coal supply model. The base coal supply file produced by the modified coal supply model was input via the coal preprocessor and the final "cookie" (final report writer results) results were very close to the base "cookie" produced with DOE's AAR77 standard coal supply table. The differences in the integrating model results are in general quite acceptable. Some question has arisen in the area of non-associated gas results. DOE personnel suggested the possibility of alternate solutions when this was discussed with them. Some differences have been found in the output of Version 1 of MAGEN but this is not of major concern. A copy of TAMU's and DOE's "cookie" is included in the final report for comparison.

The major difficulty of implementation has been the incompatibility of the MVS operating system used at DOE versus the SVS system used at Texas A&M University. Many job streams had to be broken down into multiple jobs in order to prevent various queues from overflowing. Secondary difficulties were different physical intermediate I/0 devices, lack of DASD allocation, required in or out specifications on the label parameter, and the inability to rewind DD * data files.

Many scenarios were run at TAMU in order to test the sensitivity of the overall model to major changes in selected parameters. The demand model was analyzed by initially varying the elasticities produced from the Regional Demand Forecasting Model. The consumption levels for fuels changed very little; consequently DOE was contacted for help in suggesting an appropriate parameter variation. The initial suggestion was to vary the beta price coefficient in the parameter preprocessor. A 25% beta variation was performed. The parameter file for each variation was input to the Regional Demand Forecasting System which in turn produced three files. These three files were input to the final phase of the integrating model. The consumption levels changed less than 2% for the 25% beta variation. At this point, DOE suggested a price path variation. The implication is that price paths which are input to the model (predetermined) must be changed to significantly affect demand. This result was not clear at the outset.

Variations were performed on key parameters of the oil and gas supply models including (1) finding rates, (2) discount rates, (3) price paths, and (4) rig life. The variations were completed satisfactorily to allow analysis of the importance of these parameters on oil and gas production response behavior of the models.

Variations were also done in the "mine life" parameter of the coal supply model. These variations were also completed satisfactorily allowing analysis of coal production response behavior of the model.

The oil, gas, coal, and demand model variations were interfaced pairwise with the integrating model in order to test the sensitivity of the overall market equilibrium solutions to the parameter variations.

One case (low finding rate for oil and gas) produced an unbounded solution problem. This occurred when attempting to insert the oil and gas supply curves corresponding to the 50% low finding rates. DOE was sent a magnetic tape which included the MPSX representation of the problem and a starting basis. Approximately two weeks after the reel was sent, a tentative diagnosis was received from DOE. The low production of oil and gas had resulted in no butane being produced in California. A cost of zero resulted for butane. Ultimately a conversion took place from butane to gasoline and its distillates. Since butane was free, an infinite conversion process took place. The point to be made here was that without the aid of DOE, this problem would have required an enormous amount of time to solve; one must thoroughly understand the workings of the model if such problems are to be solved. This was a new problem never encountered by DOE implying the need for more verification tests by DOE or others using the model.

The system is definitely not of the "push button" variety and requires great attention to detail in the preparation of each scenario. However, the overall operation is quite straightforward and should be mastered by most reasonably proficient systems analysts.

EFFECTS OF RESOURCE ESTIMATES AND RATES OF FINDING ON PROJECTION OF CRUDE OIL AND NATURAL GAS PRODUCTION

by W. L. Fisher, R. A. Morton and W. E. Galloway

Bureau of Economic Geology
The University of Texas at Austin

EXECUTIVE SUMMARY

A fundamental factor in the projection of future oil and gas production levels lies in the size of the reserve and resource base and degrees of certainty regarding size and distribution of that base. DOE/EIA Projection Series C (median case) calls for reserve additions through 1990 of 34 billion barrels of crude oil and 203 TCF of natural gas to support a 1985 production of 8.5 mbd of oil and 16.7 TCF of gas, and a 1990 production of 8.0 mbd of crude oil and 16.2 TCF of gas. Remaining reserves and resources are classed in three main categories: (1) measured reserves (about 30 billion barrels of oil and 205 TCF of gas), support current production and are known with a high degree of certainty; (2) inferred and indicated reserves (28 billion barrels of oil and 201 TCF of gas) are known with less, but still a reasonable degree of certainty; and (3) undiscovered, recoverable resources, which are estimated but not known. It is volumes from the second and third categories that must move to the reserve category to support future production. Resource estimates from several independent sources (U.S. Geological Survey, major oil companies, National Academy of Sciences) are in general agreement, considering inherent uncertainty of such estimates, and range from 88 to 118 billion barrels of oil and 443 to 973 TCF of natural gas. Reserve addition requirements under EIA's Projection Series C represent about 29 percent of the remaining potential of crude oil and 28 percent or less of the remaining potential of natural gas. Reserves volume, even given the uncertainty of undiscovered resource estimates, is not a significant constraint in short- and intermediate-range projections (through 1985); it becomes critical, however, in long-range projections (1990 and longer).

Virtually all projections of finding rate, measured as reserves added per physical increment of drilling, are based on mathematically-derived extrapolation of historical data. Rates of finding have declined over time as easier found oil and gas have been discovered and as progressively more difficult-to-locate oil and gas are explored for. Principal judgments in calculation of finding rate thus involve (1) what extent of past history should be used, (2) has past decline and hence projected future decline occurred as continuous decline or has it occurred in distinct epochs characterized by relative stability, and (3) are temporary shifts in the drilling mix (exploration versus development drilling) significant in calculations of finding rate? Data are presented that define a distinct post-1970 exploration epoch, for which finding rate has been relatively stable, and thus reliable for short-term (through 1985) projections. Analysis of regional shifts in the drilling mix show that these have very little significance in calculated finding rate. It is concluded that forward projections of reserve additions should utilize a finding rate in the range of 17 to 18 barrels of oil and gas equivalent per foot rather than the 21.5 boe per foot average implied in the National Energy Plan projections or the higher 24.7 boe per foot average utilized in EIA's Projection Series C through 1990. The DOE calculation of finding rate, applying a continuous decline function to the past twenty years of historical data, yields projected reserve additions 20 to 40 percent higher than recent history of finding rates would seem to warrant, and thereby underestimate the drilling effort necessary to support given production levels.

THE INVESTMENT PROCESS IN THE DOE/MEFS OIL AND GAS SUPPLY MODELS: ANALYSIS AND IMPLICATIONS

by Peter S. Rose

Finance Department
College of Business Administration
Texas A&M University

EXECUTIVE SUMMARY

The objective of this study is to examine the process of investment determination as specified in the MEFS Oil and Gas Supply Models. The study is divided into five parts: (1) a detailed examination of the MEFS Oil and Gas Supply Models; (2) an analysis of the recent economics literature dealing with the determinants of investment spending; (3) an analysis of capital budgeting theory and risk measurement in the current financial analysis literature; (4) a study of investment decision processes currently employed by petroleum firms; and (5) an analysis of recent financial trends in petroleum in an attempt to assess their implications for oil and gas investment modeling.

Investment in additional petroleum drilling is determined in the model by the net present value (NPV) technique. Each potential level of drilling activity is tested against the NPV standard to determine if it is economically viable, assuming a minimum required real rate of return of 8 percent. Future net cash flows are discounted back to their present values and compared against relevant drilling costs. Net cash flow is the difference between gross cash flow and annual operating and overhead costs. Gross cash flow is dependent upon the quantity of output expected multiplied by the assumed price of output less any royalty payments and taxes. The minimum acceptable price set for drilling is that price level which forces NPV to zero.

Revenue flows are assumed to begin in year one, except for offshore regions, and project life is assumed constant for all regions. Across a single region all costs, taxes, royalties, and the appropriate risk-adjusted discount rate are presumed constant, both regionally and over time. The models assume the absence of constraints in the availability of investment capital.

DOE modeling of the petroleum investment process suffers from lack of clarity. Documentation is incomplete and appears to be seriously out of date. The derivation of cost estimates and relevant parameters is not clear. Current documentation is so inadequate as to make it virtually impossible to determine precisely what has been added or modified by FEA and DOE personnel.

A particularly serious problem is the apparent inconsistency of the investment model with capital budgeting theory and recent developments in risk measurement. The NPV model was designed for the evaluation of a single investment project at the margin, not for determining regional drilling activity from average price and cost estimates. The regional approach averages out differences among individual firms and capital projects. This appears to introduce a bias in drilling activity at both high and low marginal prices.

The rate of discount assumed to motivate investment in the model appears to be seriously underestimated. A real discount rate of only 8 percent probably substantially underestimates the risk inherent in petroleum operations. A survey of petroleum companies suggests that a real discount rate of at least 14 to 15 percent after taxes may be more appropriate. It is important to reestimate the model using a variety of discount rates to examine the consequences of changes in the cost of capital for the volume of drilling activity undertaken.

Alternative specifications of the investment process should be tested. One potentially useful approach is to review models of investment behavior appearing in the recent econometric literature. These models have generally relied upon two sets of variables in forecasting changes in investment spending—one set related to the demand for capital goods and the other related to the marginal capital supply function. Investment demand variables include lagged changes in output and the level of real capital. Capital supply variables typically include the real rate of interest and net cash flow (principally net earnings and depreciation or the debt-equity ratio).

Probably the most serious shortcoming of the model from the standpoint of financial theory is its failure to adequately incorporate risk. For example, the model assumes that acceptance of an investment project does not change the total business-risk complexion of the firm. The NPV method provides no information on the dispersion of estimated cash flows. Cash flow estimates are not subject to probability distributions. Rather, single NPV estimates are made period by period without recognition of the possibility of variation around those estimates. To the extent that investment projects adopted add significantly to the total risk complexion of the firm, the model would tend to overestimate the volume of petroleum investment spending.

Before attempting to model industry investment decision-making, it is important to survey petroleum investment practices. In this study a survey was conducted of two populations within the petroleum sector—all integrated companies (focusing upon their drilling subsidiaries) and a selected sample of independent companies actively engaged in drilling activity. Preliminary results suggest it is inappropriate to assume that all petroleum firms use the NPV approach in making their investment decisions. A variety of approaches are used in practice and there is evidence that petroleum firms make use of NPV, pay-back, internal-rate-of-return, and accounting-based methods in combination.

The survey results also suggest the critical importance of marginal prices on new output in determining petroleum investment. Changes in marginal prices influence the net cash flow from a project and therefore its loan potential. In contrast, net cash flow affects the marginal cost of capital for petroleum firms which influences their investment. Revenue generated by current investment directly affects industry sources of financing through an impact on the availability of internally generated funds. This circular relationship between cash flows, cost of capital, and investment suggests the necessity for a multi-equation framework within the model to more realistically reflect petroleum's capital-budgeting process.

Recent profitability and risk trends within the petroleum sector and recent changes in the composition of funds uses and funds sources for the industry are explored in this study. In general, it is difficult to detect any substantial increases in the volatility (risk) of petroleum profits. However, the industry's increasing use of debt has generated the potential for greater volatility in returns to its stockholders. This factor would suggest either that the discount rate applied to expected cash streams will have to be increased or the variability of estimated cash flows (risk) included formally in the model if it is to more accurately mirror petroleum investment decisions.

CRUDE OIL PRODUCTION FROM STRIPPER WELLS

by Robert R. Hill

College of Business Administration
Texas A&M University

EXECUTIVE SUMMARY

This study evaluates the DOE oil and gas supply models' usefulness for projecting the response of the stripper well segment of the U.S. domestic oil producing industry. The study also initiates a primary data gathering process for the stripper well industry.

A stripper well property is an economically marginal oil producing property. A property is currently defined as the right to produce oil from a distinct reservoir. A stripper property, loosely speaking, is a property capable of producing 10 or less barrels of oil per well per day.

Stripper well producers perceive themselves as a distinct segment of the oil producing sector of the U.S. Further, the producers are currently treated differently from the rest of the oil producing sector by both state and national laws and regulations. This portion of the U.S. oil producers is large in numbers [369,000 wells or about 73% of total U.S. oil wells] but relatively small in contribution [392 x 10^6 bbl/year or about 14% of total U.S. production]. The relative importance of this segment is growing over time.

In the Midterm Energy Forecasting System (MEFS) the oil industry is treated as a monolith. The stripper well segment is aggregated into the rest of the industry. The critical abandonment-produce decision that is continually faced by the marginal producer is assumed away when the exploration drilling decision is made in the model. In short MEFS does not appear to model the economic characteristics of the stripper well industry.

A primary difficulty in modeling stripper well activities is the lack of necessary data. Adequate data exists on productivity and certain physical characteristics but the data is expensive to access. Data on costs and on the relationships between costs and productivity and physical characteristics do not seem to exist. This research includes the design and administration to a sample of stripper well producers of a survey to obtain cost data and correlations of the cost data to physical and productivity characteristics.

EVALUATION OF DOE/MEFS OIL AND GAS SUPPLY MODELS: BEHAVIOR OF THE COMPUTER MODEL

by John C. Stone and Russell G. Thompson

The Energy Laboratory
University of Houston

EXECUTIVE SUMMARY

The present summary pertains to an evaluation of the model of oil and gas supply developed and used for policy analysis by the Federal Energy Administration and, later, the Energy Information Administration of the Department of Energy. Like the main report, this summary is organized around the four criteria for model evaluation employed in the TNEMP study (clarity, workability, validity, verifiability) and also includes some conclusions derived from operational experience with the model.

Clarity

Once discovered, the basic conceptual framework of the model is reasonably straightforward. In implementation, however, the model is massive, complex, and even convoluted. This situation, combined with incomplete (and occasionally erroneous) documentation, renders comprehension of the model unnecessarily difficult.

Workability

From the user's viewpoint, the model is designed to answer a rather narrowly defined set of questions about the price-responsiveness of oil and gas production (at the extensive margin) under given specifications for finding rate relationships, drilling response patterns, discount rates, and other economic and technical parameters. It has principally been used to provide "supply functions" for a partial equilibrium model of the energy system, and for that purpose its general design is arguably consistent. It must be recognized, however, that the supply model is dynamic and that its interface with a static equilibrium model requires some strong assumptions about (or ignoring of) market conditions in years preceding and succeeding the target year. Because of the limited purposes of its development, the model (and modelers) would be strained to answer more detailed questions about petroleum industry financial behavior or to investigate the impact of alternative, complex regulatory measures, particularly those of a non-economic nature. These remarks are not intended as a criticism of the model, but merely an explicit recognition of its design limitations.

From the modeler's viewpoint, the model is cumbersome and difficult to operate, its efficiency and flexibility suffering greatly from the model's patchwork origins as a system of modifications to the previously existing National Petroleum Council model. The interface between its component production and drilling programs seems incomplete. Certain errors in programming and data were discovered, but their impact on model results was not great. Finally, certain questionable results from the model seem to indicate other flaws in model logic or programming.

Validity/coherence

On the whole the model attempts to be consistent with the theory of economic allocation of (drilling and production) resources, subject to declining finding rates and increasing costs of drilling. Certain assumptions, however, are not consistent with this theory, with common sense, or with the basic assumptions and theory underlying the overall partial equilibrium model of the energy system. Principal among these are:

(1) The model essentially ignores the basic interdependence of oil and gas exploration and production. By maintaining separate oil and gas programs, the model does not allow for transfer of drilling activity from one resource to the other (only independent expansion paths from a given starting point). Thus, the model ignores an important degree of freedom in the response of drilling to changes in relative prices and costs of production.

(2) In support of and required by the separate modeling of oil and gas exploration and production, the model employs *a priori* pricing of associated/dissolved gas and natural gas liquids (in proportion to the price of the primary product). This is arbitrary, results in different pricing for associated and nonassociated gas, and is in general inconsistent with the objective of determining equilibrium prices and quantities for petroleum products.

(3) The model assumes pre-determined project lives and depletion rates for production from oil and gas fields (old and new) and thus does not consider the price-responsiveness of production at the intensive margin.

Verifiability/correspondence

The greatest strength of the model is probably in its region-specific data base of economic and technical parameters. This data base is quite large and has not been rigorously examined, but it seems unlikely that an unambiguously superior data base could be readily assimilated in the public sector against which to check that employed in the model.

Aside from parameter estimates, the model requires specification of various behavioral and technical relationships, three of which are most critical.

(1) The model employs finding rate relationships based on a long-run historical data and a functional form which implies continuously decreasing finding rates. These specifications have apparently become somewhat controversial. Independent of the unresolved question of the general "behavior" of finding rates, the relationships employed in the model suffer from two unfortunate constraints on their specification. First, the same functional form is applied to each region, which is unnecessary and detracts from the fit to historical data. Second, the relationships are constrained to approach asymptotically the USGS estimates of undiscovered resources. As there is no way to assure even rough consistency between these estimates and the economic specifications for use of the model, this constraint seems inappropriate at best, particularly since it causes substantial departures from the pattern indicated by historical data.

(2) Given the pre-determined depletion rate specifications, the pattern of new drilling is the only source of price sensitivity captured by the model. Unfortunately, the relationships governing the inter-temporal and inter-regional allocation of drilling activity are probably the most arbitrary behavioral specifications employed in the model. While the allocation logic applied is plausible enough (and internally consistent in principle), it is by no means the only plausible logic and seems to be based more on heuristics than on any particular theory of economic behavior.

(3) A net present value calculation is used to determine profitability of production at specified prices. While net present value decision-making may not be a widely applicable specification for individual firms, it may be acceptable as a representation of industry behavior in the aggregrate. (There is no unambiguously superior specification.) In implementation, however, data limitations have forced the use of average drilling and production cost estimates in what is properly a marginal calculation. This will bias supply estimates in a direction dependent upon the relative positions of the average and marginal cost curves at the specified prices. A specification of eight percent as a real rate of return after tax is also arguably low.

Operational experience

As indicated previously, experience with the model in Texas has shown it to be cumbersome and difficult to operate. A particular effort was directed to obtaining a series of runs at constant price trajectories so as to obtain measurable and compatible estimates of the sensitivity of oil and gas supply projections to price, finding rate, discount rate, and drilling rate (measured as the fraction of profitable drilling completed by a given year). The results of these analyses may be summarized as follows:

(1) The pattern of offshore production for both oil and gas is somewhat erratic, and this behavior would seem to indicate the existence of further logic or programming errors in the drilling allocation program. In light of this situation, further analysis of the offshore results would not be viable.

(2) The pattern of onshore production response is (with trivial exceptions) consistent with prior expectations. It is not known, however, to what extent the magnitudes of observed results might be affected by the problems creating the erratic offshore results.

(3) Arc elasticities of supply were calculated anyway for the onshore results in years 1985 and 1990. It is difficult to generalize about the sensitivity results, except to note a tendency of all elasticities to decrease with increasing price. Some computed elasticities are quite large because of changes to or from zero production levels. Price elasticities are highly sensitive to the specifications for finding rate, discount rate, and drilling rate; those for gas range between 0.2 and 3.7, while those for oil range between 0 and 15. Supply projections are generally more sensitive to finding rate than to discount rate, especially at higher prices. The impact of the drilling rate is generally less than that of the finding rate but frequently exceeds that of the discount rate at higher prices. The general ranges of these elasticities are tabulated below.

"Variable"	GAS	OIL
Finding rate	1.1 to 3.0	1.6 to 3.0
Discount rate	-0.2 to -3.3	-0.5 to -3.3
Drilling rate	0.5 to 1.7	0 to 1

On the future of the model

The present evaluation indicates a serious need for replacement of the current model with a sound economic model for policy evaluation. Reportedly, DOE/EIA awarded a contract in September 1978 (to the contractor involved with development of the present model for use by FEA and EIA) for the development of a new model of oil and gas supply.

EVALUATION OF THE DOE/MEFS COAL SUPPLY MODEL: BEHAVIOR OF THE COMPUTER MODEL

by F. D. Singleton and Russell G. Thompson

The Energy Laboratory
University of Houston

EXECUTIVE SUMMARY

Results of the coal supply model evaluation can be discussed under the following headings:

Validity/coherence

The coal supply model is a valid representation of coal supplies in the intermediate term for which it was designed. Supplies of various types of coal are determined in twelve regions as a function of the marginal price. However, in the long term, i.e., beyond 1990, the model has no provision for retiring "new" mines to account for the dynamics of coal depletion. Also, in the long term the economic reserves of coal will be expanded by exploration, but the model does not take this into account. (Actually, the lignite reserves in Texas have been expanded already since the development of the model.) In the short term, the model adjusts faster than is possible in the industry because lag times between build decision and production start-up are not explicitly taken into account.

Verifiability/correspondence

The model includes the entire Demonstrated Reserves data base, as developed by the U. S. Bureau of Mines. Generally no numbers differ widely (i.e., by orders of magnitude) from values in the literature on coal. However, the mine lifetime should be allowed to be different for different types of mines and different regions. The capital costs should be escalated to reflect recent price indices. Btu losses due to coal cleaning are probably too high in certain cases.

Clarity

Documentation of the coal supply model is reasonably adequate since the basic ideas have remained unchanged since the inception of PIES. Financial considerations and computer algorithms are poorly described. (Laborious dissection of the programming code is necessary to discern details.) Improved clarity has been achieved at the University of Houston by eliminating one of the three computer languages involved.

Workability

The model is believed to be workable for its intended purpose of defining coal supply curves for 1985 and 1990 solutions of the PIES/MEFS Integrating Model. Since much regional and coal quality detail is present, the model is expected to be adequate to indicate both regional and national trends in the coal industry (provided that the transportation, electric utility, and other coal using modules in PIES/MEFS are adequate). The supply curves are expected to lose precision as the coal quality parameter is increased due to the various model weaknesses mentioned above.

Model operating experience

Trial runs of the FORTRAN portions of the model were first made at the University of Houston and a FORTRAN package was developed to replace the GAMMA language portion of the model. The FORTRAN modules were delivered to Texas A&M University where they were used to generate a complete model from scratch. This generation replicated the 1985 Base Case Integrating Model Solution in the Administrator's Annual Report. The effect of increasing the mine life from 20 years to 30 years was investigated in a second complete model run. Increasing the mine life parameter rotated the supply curves counterclockwise about their origin. This results in a major increase in price for very large quantities, but for smaller quantities near the price axis the price increase is smaller. (Actually depreciation cost is lowered so there is a slight decrease in cost near the price axis). Increasing the mine life from 20 to 30 years increased the slopes of the supply curves in 1985 by a factor of 1.5, but this did not significantly affect the Integrating Model solution because most levels of coal supplies were relatively small in comparison to their supply potential. Coal prices typically increased by about five percent (with some actually decreasing slightly).

Several programming errors (and illogical program features) were found in the model. None of these problems significantly affected the supply curves and results of the integrating model solution. However, some values in the reported output tables of mine capital costs were computed incorrectly.

EVALUATION OF STRUCTURE AND ASSUMPTIONS OF THE DOE/MEFS PETROLEUM REFINERY AND SYNTHETIC FUELS MODELS

by Frank A. Taylor and Russell G. Thompson

The Energy Laboratory
University of Houston

EXECUTIVE SUMMARY

Several highlights stand out from the evaluation of the MEFS (PIES) petroleum refining model, as used by the U. S. Department of Energy. These highlights are as follows:

- The PIES petroleum refinery model is a mathematical representation of a mathematical representation, with the original representation being proprietary. Technical structural delineations are totally lost in developing a single vector to represent an entire refinery. Single vector representations simplify computational considerations in the PIES integrating model, but this simplicity largely sacrifices the capability to analyze the consequences of structural change induced by policy. Further, identification of an explicit technological representation of the refining industry from the solution of the model is impossible if a linear combination of two or more vectors is optimal. This limitation seems to foreclose completely the possibility of identifying from the equilibrium solutions the underlying economic supplies of refinery products and derived economic demands for refinery inputs; see J. Marshak's introductory article in *Statistical Inference in Dynamic Economic Models.* If identification of the underlying economic demands and supplies is impossible, then the model lacks capability to clearly identify policy issues that require sound economic inputs. Thus, the model may not have any capability to assist policy-makers in resolving significant policy issues. As defined for this study, the model fails the test of workability.
- Analysis of a representative regional module of the refinery model showed that the simplified model only produced reasonable results within certain ranges of parametric variations. If these ranges are exceeded, then the solutions of the model are erratic and seemingly unreliable. This limitation means that the evaluations made with the

PIES model must be limited to those that *a priori* will not exceed these ranges of reliability. Outside its reliable range the model could not be used to identify or resolve issues. Nor would it offer results applicable in any practical way to analysis of the user's problems, and therefore, outside a certain range, the model can be said to fail the workability test in its actual outputs as well as in its theoretical basis as outlined above. Since erratic results are obviously not part of the modeler's expectations the model can be said to fail the test of verifiability outside its reliable range.

- The structure of the PIES refinery model offers some grounds for criticism of the model's soundness; questionable assumptions include the ability to achieve a given output profile at a given cost regardless of crude type and optimistic blending assumptions. Single vector refinery representations give rise to rigidities in the model not found in the actual refining industry. These rigidities and lack of environmental detail raise the possibility that the model might misrepresent the regional impact on the refining industry of various policy decisions or external stimuli. Also, coal liquefaction technology representations do not seem to have product output profiles that are similar to those of coal liquefaction processes documented in the published literature. These concerns relate to the criteria of validity and workability.
- Because the structure of the model is simple, its clarity is good. Its purposes and interactions are reasonably clear to a modeler, and its outputs should be clear to laymen with a reasonable knowledge of the general subject matter. The documentation is adequate for the most part, but it is not clear how the specific coefficients were derived.

THE REPRESENTATION OF THE ELECTRIC UTILITY SECTOR IN THE DOE/MEFS: OVERVIEW AND COMMENT

by Lawrence K. Vanston and Martin L. Baughman

EXECUTIVE SUMMARY

In light of the evaluation criteria, the study makes three observations. First, the electric utility representation in the LP is the simplest possible. The only additional simplifications that could be made are to reduce the number of plant types or to reduce the number of load categories. Either would sacrifice appropriateness as an electric utility representation.

Second, the model will forecast behavior very poorly if unconstrained. However, the model has been used for the purpose of accounting; not for determination of behavior. Future expansion plans and plant utilitization possibilities are entered as a very tightly constrained set of data. The model then is used simply to translate one set of prices (fuel and capital) into another (electricity).

Third, as an analytical tool in and of its own right, the model offers nothing that could not be obtained in more detail elsewhere. Thus we conclude that the electricity sector of the PIES model serves its intended, although somewhat limited, purpose adequately.

THE TRANSPORTATION MODEL IN THE DOE/MEFS: DOCUMENTATION AND CRITIQUE

by William F. McFarland and Carol Bradley

Texas Transportation Institute
Texas A&M University

EXECUTIVE SUMMARY

General Description of Transportation Models

Pipelines

Three basic methods are used to estimate the costs of transporting energy products by pipeline:

(1) Transportation cost = Centroid-to-Centroid Tariff + Gathering Charges (if any)

(2) Transportation Cost = (Rate Per Mile × Centroid-to-Centroid distance) + Gathering Charges (if any)

(3) Rate of Return formula used to estimate annual pipeline cost which is divided by annual pipeline throughput to derive transport cost per barrel.

Method (1) is used to estimate transportation costs for crude oil and petroleum products in existing pipeline. Method (2) is used for crude oil, syncrude, and petroleum products where existing tariffs are not available, and also is used for all natural gas pipelines. In Method (2) the "rate per mile" depends on the type of product and the year. In Methods (1) and (2) gathering charges are usually based on existing tariffs or are assumed values that depend on whether production is from established fields and is off-shore or onshore. There are no gathering charges for petroleum products and residuals. Method (3) is used to estimate the transportation cost of Alaskan crude and natural gas.

Tankers

The cost of transporting crude, petroleum products, and residuals by tanker is estimated from a cost formula that considers average operating costs and operating characteristics for two different sizes of tankers, one for crude and another for products and residuals.

Barges

The cost of transporting petroleum products and residuals by barge is estimated from average rates observed for operating barges on inland waterways. Cost data from a previous research report are updated and used in the mid-1977 version of MEFS.

Coal by Rail and Barge

Coal transport costs are estimated using a detailed rail-barge transshipment network consisting of many linked nodes. Costs of transporting coal by rail are estimated from a linear regression equation based on unit train rates in the Midwest, giving a linear relation between cost per ton mile and distance of haul. Adjustments to these rates, based on "rules of thumb," are made for costs in the Northeast Corridor and over the Rocky Mountains.

In the coal transshipment network, barge costs are estimated from weighted average rates observed on different waterways.

Critique of Transportation Models

The regional supply-demand structure of MEFS using centroids within regions to represent the weighted centers of activity of different types, is an excellent method to use in modeling energy transportation in the United States. However, several of the regions, such as the refinery region for the south-southwestern U.S. and most demand regions, probably are too large.

Another major shortcoming of the MEFS transportation models is that all transport costs are estimated in 1975 dollars, and it simply is assumed that costs in 1980, 1985, and 1990 remain constant in real 1975 dollars. This assumption probably overstates the cost of transporting by pipeline especially in the later years of the forecast. To be valid, the forecasting method should have considered explicitly the trends in real transport cost by each mode over time.

The transportation models used in MEFS have not been verified either *indirectly,* through verifying the overall MEFS model or energy flows, or directly, through determining how well the models predict transportation costs for different material/mode combinations. The models probably are best at predicting pipeline costs for existing pipelines for 1980, and are less accurate in predicting for later years and new pipelines. The coal model is considerably weakened by use of a regression equation based on only a partial sample of U.S. train rates which does not consider trends in rates over time.

Tanker rates used in MEFS also have not been verified but probably are sufficiently accurate that they do not substantially distort the overall model solution, given their relative insignificance in the overall transport model.

Although the concepts used in the transportation models in MEFS are easily understood, many of the data inputs are inadequately documented, which considerably limits the full understanding of these models by users outside of the U.S. Department of Energy. This lack of complete documentation, together with the lack of verification, either of the overall MEFS model or the specific transportation models, leads us to conclude that the MEFS model cannot be relied upon by outside users at this time to evaluate important energy policies. We conclude that high priority should be given to structuring the model in such a way that it can be verified (validated) using data from recent years. Also, the transportation models should be reformulated to: (1) explicitly consider trends in transport costs by mode, (2) consider alternative methods of modeling coal transport among regions, possibly using additional demand nodes, and (3) use more definitive models for estimating costs of new or upgraded pipelines and rail lines.

EVALUATION OF DOE/MEFS DEMAND MODEL

by Bob Davis and Hong Y. Lee

Department of Agricultural Economics
Texas Tech University

EXECUTIVE SUMMARY

The objective of this study is to examine the demand model of the Midterm Energy Forecasting System (MEFS) of the Department of Energy. The demand model consists of three major sectors — residential, commercial and industrial sector, transportation sector and minor fuels sector. Four preprocessors provide inputs into the demand model, outputs from which in turn are fed into the supply-demand integrating model to derive a set of equilibrium prices and quantities by types of fuel, by sectors, and by regions of the U.S. The demand model is econometric and utilizes both two-stage estimation process and single equation ordinary least squares estimation procedures. The reliability of estimates of most parameters are not evaluated due to lack of adequate test statistics. The results of sensitivity analyses seem to indicate that the model is highly stable in view of relatively large modifications employed in the analyses.

The price perturbations employed in converting the dynamic econometric model results into a static representation of demand for use in the integrating model may have led to incorrect specification of the static demand functions because of the quantity dependence on lagged quantity, which in turn is a function of lagged price. This may partially explain the insensitivity of the entire demand model to relatively large modifications in elasticities and other parameter estimates.

Other explanations on the insensitivity of the demand model to relatively large changes may be (1) the severity of the constraints placed on the model and/or (2) the influence of non-price variables in the model. In view of the insensitivity of the model energy policies intended to affect consumption through prices may not be correctly identified by the model.

Other demand studies reported in the literature, while perhaps smaller than the MEFS model, have successfully employed simultaneous systems as the primary estimation procedure. Thus, it may be advisable that the researchers at the Energy Information Administration of the DOE re-examine the two stage estimation procedure in favor of the simultaneous systems approach provided adequate quality data are available. Proper use of the simultaneous systems approach would ensure that the demand model would be adequately specified and identified.

While much documentation on the demand model is available, it was not adequate to allow an in-depth, comprehensive review of the model. Many details were either omitted from the documentation or simply not available.

In summary, the MEFS demand model may have validity problems in terms of specification and estimation procedures employed. The workability of the model is questionable if it is to be used to examine price impacts on energy consumption. Clarity could be improved through more complete documentation. More complete documentation would have also made the verifiability of the model easier.

MACROECONOMIC/MICROECONOMIC INTERFACE

by James Jonish and Theodore Taylor

Department of Economics
Texas Tech University

EXECUTIVE SUMMARY

The purpose of this study is to examine the linkages between the macroeconomic and energy models utilized by the Energy Information Administration of the Department of Energy in the preparation of recent annual national energy reports. In addition, the linkages between the macroeconomic (DRI)-energy (MEFS) models and regional and international satellite models are examined. The strengths and limitations of these model linkages are critically evaluated from theoretical, empirical and policy perspectives. The limitations in these model interactions suggest that caution be exercised in utilizing the interactive model system's output for policy analysis.

The results of this study can be summarized briefly in four major areas. These are:

Macroeconomic (DRI)-Energy (MEFS) Interface

Two basic findings are of importance here. First, the structure of the linkage between the DRI macro model and the MEFS energy model is seriously inadequate. Since the two models cannot be made simultaneous, decisions have had to be made regarding the extent of the linkage between the two models. Much relevant and useful information is lost in making the linkage from the MEFS energy model to the DRI macroeconomic model, including all regional and sectoral detail. Only an aggregate of energy prices (the wholesale price index for fuels) and two detailed prices and quantities are transmitted to the macroeconomic model via the energy sector.

Many of the adjustments to the models that are done prior to their interfacing are judgmental, being based only on the off-line or prior evaluations by modelers at Data Resources or the Department of Energy. This feature makes evaluation of the efficacy of the models very difficult.

Second, in examining various model results it has been found that at the national level changes in the energy sector have, over the long run, very little impact on production or prices. Reductions in total output will be on the order of one-tenth those of reductions in energy input. The relationship between energy use and capital use does indicate that higher prices of energy will result in lower output, lower rates of capital accumulation, and lower rates of economic growth, but these effects are small relative to the changes in the price of energy. There may be substantial regional, sectoral, or industrial effects of changing energy prices, but these effects are not readily discernible by the models, due to their inflexible input-output structures. There may be considerable short-run dislocations at the national level that are not captured by the long-run equilibrium solutions provided by the DRI-MEFS model outputs.

Energy-Regional Economy Interface

The Department of Energy uses an input-output table compiled by the Bureau of Labor Statistics and a bridge table to construct regional production and employment indices from aggregate output of the DRI macroeconomic model. These structures are inflexible and do not permit the use of any locational variables. Consequently, for a particular period of time, and over time, regions develop in lock step with the national economy. This system is too simplified to provide much meaningful regional analysis to policy makers concerned with the regional costs and benefits of alternative energy policies.

Moreover, the PIES regional system determines labor requirements for each region, and implicitly assumes that the quantity of labor demanded in each region will be equal to the quantity of labor supplied. Labor supply is not modeled formally, nor are other resource constraints.

Energy-International Economy Interface

Attempts to evaluate the relationship between the national economy and other economies have been undertaken. Functional relations for oil-exporting as well as oil-importing countries are modeled, as are the domestic production impacts of changes in international oil supplies. However, these efforts suffer several serious shortcomings. Exchange rates and OPEC absorption rates are not modeled, or they are exogenous to the models. The relatively new concepts of feedback elasticity and carry-forward rate are assumed to have predetermined values, but might be econometrically modeled. These limitations cause much of the analysis to be judgmental, with specific results being dependent on predetermined values of critical variables.

INDEX

D

E

F

G

H

I

J

K

L

M

N

T

U

V